About Island Press

Island Press is the only nonprofit organization in the United States whose principal purpose is the publication of books on environmental issues and natural resource management. We provide solutions-oriented information to professionals, public officials, business and community leaders, and concerned citizens who are shaping responses to environmental problems.

In 2004, Island Press celebrates its twentieth anniversary as the leading provider of timely and practical books that take a multidisciplinary approach to critical environmental concerns. Our growing list of titles reflects our commitment to bringing the best of an expanding body of literature to the environmental community throughout North America and the world.

Support for Island Press is provided by the Agua Fund, Brainerd Foundation, Geraldine R. Dodge Foundation, Doris Duke Charitable Foundation, Educational Foundation of America, The Ford Foundation, The George Gund Foundation, The William and Flora Hewlett Foundation, Henry Luce Foundation, The John D. and Catherine T. MacArthur Foundation, The Andrew W. Mellon Foundation, The Curtis and Edith Munson Foundation, National Environmental Trust, The New-Land Foundation, Oak Foundation, The Overbrook Foundation, The David and Lucile Packard Foundation, The Pew Charitable Trusts, The Rockefeller Foundation, The Winslow Foundation, and other generous donors.

The opinions expressed in this book are those of the author(s) and do not necessarily reflect the views of these foundations.

About SCOPE

The Scientific Committee on Problems of the Environment (SCOPE) was established by the International Council for Science (ICSU) in 1969. It brings together natural and social scientists to identify emerging or potential environmental issues and to address jointly the nature and solution of environmental problems on a global basis. Operating at an interface between the science and decision-making sectors, SCOPE's interdisciplinary and critical focus on available knowledge provides analytical and practical tools to promote further research and more sustainable management of the Earth's resources. SCOPE's members, forty national science academies and research councils and twenty-two international scientific unions, committees, and societies, guide and develop its scientific program.

SCOPE 65

Agriculture and the Nitrogen Cycle

The Scientific Committee on Problems of the Environment (SCOPE)

SCOPE Series

SCOPE 1–59 in the series were published by John Wiley & Sons, Ltd., U.K. Island Press is the publisher for SCOPE 60 as well as subsequent titles in the series.

SCOPE 60: *Resilience and the Behavior of Large-Scale Systems,* edited by Lance H. Gunderson and Lowell Pritchard Jr.

SCOPE 61: *Interactions of the Major Biogeochemical Cycles: Global Change and Human Impacts,* edited by Jerry M. Melillo, Christopher B. Field, and Bedrich Moldan

SCOPE 62: *The Global Carbon Cycle: Integrating Humans, Climate, and the Natural World,* edited by Christopher B. Field and Michael R. Raupach

SCOPE 63: *Invasive Alien Species: A New Synthesis,* edited by Harold A. Mooney et al.

SCOPE 64: *Sustaining Biodiversity and Ecosystem Services in Soils and Sediments,* edited by Diana H. Wall

SCOPE 65: *Agriculture and the Nitrogen Cycle: Assessing the Impacts of Fertilizer Use on Food Production and the Environment,* edited by Arvin R. Mosier, J. Keith Syers, and John R. Freney

SCOPE 65

Agriculture and the Nitrogen Cycle

Assessing the Impacts of Fertilizer Use on Food Production and the Environment

Edited by
**Arvin R. Mosier, J. Keith Syers,
and John R. Freney**

A project of SCOPE, the Scientific Committee on
Problems of the Environment, of the
International Council for Science

ISLAND PRESS

Washington • Covelo • London

Library of Congress Cataloging-in-Publication Data

Agriculture and the nitrogen cycle : assessing the impacts of fertilizer use on food production and the environment / edited by Arvin R. Mosier, J. Keith Syers, and John R. Freney.

    p. cm. — (SCOPE ; 65)

Includes bibliographical references and index. (p.).

ISBN 1-55963-708-0 (cloth : alk. paper) -- ISBN 1-55963-710-2 (pbk. : alk. paper)   1. Nitrogen fertilizers.   2. Nitrogen fertilizers — Environmental aspects.   3. Nitrogen cycle.   I. Mosier, Arvin R.   II. Syers, John K. (John Keith)   III. Freney, John R. (John Raymond)   IV. SCOPE report ; 65

S651.8.A32 2004

631.8'4 — dc22   2004012075

British Cataloguing-in-Publication data available.

Printed on recycled, acid-free paper

Manufactured in the United States of America
10 9 8 7 6 5 4 3 2 1

Contents

Part I: Overview

1. Nitrogen Fertilizer: An Essential Component of Increased Food, Feed and Fiber Production
Arvin R. Mosier, J. Keith Syers, and John R. Freney

Part II: Crosscutting Issues

2. Crop, Environmental, and Management Factors Affecting Nitrogen Use Efficiency
Vethaiya Balasubramanian, Bruno Alves, Milkha Aulakh, Mateete Bekunda, Zucong Cai, Laurie Drinkwater, Daniel Mugendi, Chris van Kessel, and Oene Oenema

3. Emerging Technologies to Increase the Efficiency of Use of Fertilizer Nitrogen
Ken E. Giller, Phil Chalk, Achim Dobermann, Larry Hammond, Patrick Heffer, Jagdish K. Ladha, Phibion Nyamudeza, Luc Maene, Harry Ssali, and John Freney

Part III: Low-input Systems

Part IV: High-input Systems

List of Figures and Tables

Figures

Tables

Foreword

The Scientific Committee on Problems of the Environment (SCOPE), in collaboration with the International Geosphere Biosphere Programme (IGBP), publishes this book as the third in a series of Rapid Assessments of the important biogeochemical life cycles that are essential to life on this planet. The aim of this activity is to evaluate recent advances in understanding the role of nitrogen in biochemical cycling, to assess the state of knowledge of the role of fertilizer in the nitrogen cycle, and to determine the range of possible research problems related to nitrogen-based fertilizers. The SCOPE Rapid Assessment series, in conjunction with the IGBP Fast-Track Initiative, attempts to ensure that information, so generated, is published and made available within a year from the date of the synthesis. These volumes provide timely and authoritative syntheses of important issues for scientists, students, and policy makers.

This volume's main concept is that nitrogen is essential to the survival of all life forms. Yet the natural abundance of usable nitrogen is so low that massive human alteration has been required to sustain the feeding of the world's population. These changes in the normal cycling of nitrogen have exacerbated numerous environmental issues, including climate change, coastal eutrophication, and acid deposition, all of which have impacts on people and ecosystems on a regional or global basis. Global-scale alteration of the nitrogen cycle has been of concern for more than four decades, and steady advances have been made in our understanding of natural and anthropogenic components of the nitrogen cycle. This book assesses our knowledge of the forms and amounts of fertilizer nitrogen applied by crop and region, the amount of this nitrogen used by the crop, and the fate of the unused nitrogen in the environment. Further, it examines the policies that control the demand and use of fertilizer nitrogen.

SCOPE is one of 26 interdisciplinary bodies established by the International Council of Science (ICSU) to address cross-disciplinary issues. SCOPE was established by ICSU in 1969 in response to environmental concerns emerging at that time, in recognition that many of these concerns required scientific input spanning several disciplines represented within its membership. Representatives of 40 countries and 22 international, disciplinary-specific unions, scientific committees, and associates currently participate in the work of SCOPE, which directs particular attention to developing countries.

This synthesis volume is part of a joint program of two ICSU-sponsored bodies, SCOPE and the International Geosphere Biosphere Programme (IGBP), which established the International Nitrogen Initiative (INI). The INI is organized on a regional basis to assess knowledge of nitrogen flows and problems; develop region-specific solutions; implement scientific, engineering, and policy tools to solve problems; and integrate regional assessments to create an overall global assessment.

John W. B. Stewart, Editor-in-Chief

SCOPE Secretariat
51 Boulevard de Montmorency, 75016 Paris, France
Véronique Plocq Fichelet, Executive Director

Preface

Nitrogen (N) availability is a key factor in food, feed, and fiber production. Providing plant-available N through synthetic fertilizer in the 20th and 21st centuries has contributed greatly to the increased production needed to feed and clothe the increasing human population. Because of greater accessibility to N fertilizer, human activity has greatly altered nitrogen cycling globally and at the scale of large regions.

Information about the components of the N cycle has accumulated at a rapid pace in the last decade, especially with regard to processes of transfer in different terrestrial, aquatic, and atmospheric environments. There is a need to synthesize this information and assess the effect of adding additional N to natural and cultivated ecosystems. Improvements need to be made to the currently low efficiency with which fertilizer N is used within production systems if we are to continue to meet the global demands for food, animal feed, and fiber and minimize environmental problems. Major uncertainties remain, however, about the fate of fertilizer N added to agricultural soils and the potential for reducing emissions to the environment. Enhancing the technical and economic efficiency of fertilizer N is essential for both agricultural production and protection of the environment.

SCOPE (Scientific Committee on Problems of the Environment), whose mandate has been to assemble, review, and assess the information available on human-induced environmental changes, has summarized information on the biogeochemistry of N several times since 1981 (Boyer and Howarth 2002; Clark and Rosswall 1981; Howarth 1996). SCOPE has joined forces with the IGBP (International Geosphere–Biosphere Programme) to develop the International Nitrogen Initiative (INI), which was formed following the World Summit on Sustainable Development in Johannesburg on August 29, 2002.

The goal of INI is to develop a sustainable approach to managing N and thus provide food and energy to the world while minimizing the release of reactive N compounds to the environment (reactive N is biologically, photochemically, and radiatively active forms of N compounds in the atmosphere and biosphere of the Earth). INI builds on two major international conferences on N biogeochemistry (Galloway et al. 2002; van der Hoek 1998).

This book is an international assessment of the efficiency and consequences of fertilizer N and is a first step in the development of the science base for the INI. It assesses the fate of fertilizer N in the context of overall N inputs to agricultural systems, with a view to enhancing the efficiency of N use and reducing negative impacts on the environment. The book consists of an overview synthesis paper, four papers developed from discussions of cross-cutting issues, an invited paper that assesses current knowledge about the environmental dimensions of fertilizer N, and 13 papers on various aspects of fertilizer N use. The cross-cutting issues relate to the efficiency of fertilizer N use as determined by environmental and management factors, the role of emerging technologies (e.g., genetic enhancement) on the efficiency of fertilizer N use, impacts of N loss on human health and the environment, and societal responses to meeting N needs in different regions.

SCOPE publishes this book as the third of a series of rapid assessments of environmental issues. SCOPE's aim is to make sure that experts meet on a regular basis, summarize recent advances in related disciplines, and discuss their possible significance in understanding environmental problems and potential solutions. The desire is to make this information available in published form within six to nine months of an assessment. The assessment for this book was conducted at a workshop that was held in Kampala, Uganda, in January 2004.

Arvin R. Mosier, J. Keith Syers, and John R. Freney, NFRAP Editors

Literature Cited

Boyer, E. W., and R. W. Howarth. 2002. *The nitrogen cycle at regional to global scales.* Dordrecht, The Netherlands: Kluwer Academic Publishers.

Clark, F. E. and T. Rosswell, eds. 1981 Terrestrial Nitrogen Cycles: Processes, Ecosystem Strategies, and Management Impacts. *Ecological Bulletin* 33, Stockholm.

Galloway, J. N., E. B. Cowling, S. P. Seitzinger, and R. H. Socolow. 2002. Reactive nitrogen: Too much of a good thing? *Ambio* 31:60–63.

Van der Hoek, K. W. 1998. Nitrogen efficiency in global animal production. *Environmental Pollution* 102:127–132.

Acknowledgments

For financial support for this project, SCOPE thanks the International Fertilizer Industry Association (IFA), United States Department of Agriculture–Foreign Agricultural Service (USDA–FAS), USDA–Agricultural Research Service (USDA–ARS), Global Change SysTem for Analysis, Research and Training (START), U.S. National Science Foundation (NSF), International Geosphere-Biosphere Programme (IGBP), International Nitrogen Initiative (INI), International Council for Science (ICSU) and the United Nations Educational, Scientific and Cultural Organization (UNESCO), USAID–Agricultural Productivity Enhancement Program—Uganda, and Australian Centre for International Agricultural Research (ACIAR), and A.W. Mellon Foundation.

The assessment workshop was held in Kampala, Uganda, at the Grand Imperial Hotel and hosted by Professor Mateete Bekunda, Dean of Agriculture at Makerere University. We are indebted to Professor Bekunda and his staff, who provided an excellent venue for the workshop. We extend our special thanks to Mr. Edward Businge, who ably handled our IT problems and to Ms. Susan Greenwood Etienne of the SCOPE secretariat for her work to make the Kampala workshop a success. We also thank Dr. Dork Sahagian and his staff at the IGBP–GAIM Secretariat at the University of New Hampshire for their assistance in facilitating travel for the U.S. participants for the Nitrogen Fertilizer Rapid Assessment Project (NFRAP).

An additional half-day symposium, Fertilizer Nitrogen and Crop Production in Africa, was chaired by the Honorable John Odit, Chairman of the Ugandan Parliamentary Subcommittee on Agriculture. The symposium, sponsored by IFA and organized by Professor Bekunda, was held on January 14 at the Hotel Equatoria in Kampala.

Finally, we thank David Jensen for his work in reformatting figures and Susan Crookall for manuscript proofreading.

PART I
Overview

1

Nitrogen Fertilizer: An Essential Component of Increased Food, Feed, and Fiber Production

Arvin R. Mosier, J. Keith Syers, and John R. Freney

Nitrogen (N) fertilizer has made a substantial contribution to the tripling of global food production over the past 50 years. World grain production was 631 million tons in 1950 (247 kg person^{-1}) and 1840 million tons in 2000 (303 kg person^{-1}); per capita grain production peaked in 1984 at 342 kg person^{-1}.

Since 1962 annual production of N fertilizer has increased from 13.5 to 86.4 Tg (1 Tg = 10^{12} g) N in 2001 worldwide (FAO 2004). Unfortunately, the distribution of fertilizer N use is not uniform globally; so in some areas of the world, sub-Saharan Africa (SSA), for example, little fertilizer N is used (in 2001 only 1.1 kg person^{-1} compared with 22 kg person^{-1} in China), and local food production has not kept up with the increase in human population. As a consequence the protein supply per person in SSA is only 10 g day^{-1} compared with 100 g day^{-1} for people in developed countries. The limited availability of fertilizers in SSA has contributed to the decline in soil fertility through the loss of soil organic matter (Greenland 1988; Syers 1997).

In other areas of the world (e.g., Europe), excessive fertilizer N is sometimes used. Excessive use of N can lead to numerous problems directly related to human health (e.g., respiratory diseases induced by exposure to high concentrations of ozone and fine particulate matter) and ecosystem vulnerability (e.g., acidification of soils and eutrophication of coastal systems) (Cowling et al. 2001, Boyer and Howarth 2002, Galloway et al. 2002b, Mosier et al. 2002).

Little new land is suitable for crop production; therefore, the output per unit area must increase to meet an expected world population of 8.9 billion people by 2050 (FAO 2004). If the efficiency of nitrogen use (NUE) is not improved, marginal lands, including those on steep slopes, will be brought into production to help meet rising food needs, and the result will be increasing land degradation. Because of the limitation on arable land area and the need to minimize the pollution of waters and the atmosphere, the efficiency of the use of fertilizer N must be improved to sustain land quality to feed the growing population (Cassman et al. 2002).

Global Nitrogen Fertilizer Consumption

The global demand for N fertilizer is dictated largely by cereal grain production (Cassman et al. 2002). From 1995 to 1997, about 65 percent of the global N fertilizer consumed was for producing cereal grains (IFA/FAO 2001). IFA and FAO project that the relative amount of N fertilizer used by 2015 will remain unchanged but that total N consumption in cereal production will increase by about 15 percent. The increased demand for cereal production, and thus N fertilizer, is fueled mainly by human population growth but also by increased consumption of animal products on a per capita basis (Boyer et al., Chapter 16; Roy and Hammond, Chapter 17; Wood et al., Chapter 18, this volume). During the 40 years between 1961 and 2001, the human population of the world doubled from 3078 to 6134 million persons (FAO, 2004); grain production, meat production, and N fertilizer consumption, however, increased by 140, 230, and 600 percent, respectively. On a per capita basis, the respective increases were 21, 67, and 254 percent during this period.

Fertilizer N has contributed an estimated 40 percent to the increases in per capita food production over the past 50 years (Brown 1999; Smil 2002). This global figure does not reflect local and regional differences in food supply and demand. It also does not reflect the varying efficiencies of fertilizer N use in crop production across regions. For example, in 2001, on a per capita basis, N fertilizer consumption in the United States was 38 kg person^{-1}, 11 kg person^{-1} in India, but only 1.1 kg person^{-1} in SSA. There are a variety of reasons for the inequities in fertilizer N distribution around the globe. In some parts of Asia, Europe, and North America, fertilizer is relatively inexpensive and available to farmers. In SSA and in parts of Asia, the cost is high (as much as five times the global market price; Roy and Hammond, Chapter 17, this volume) and supply is limited.

As a result of the high cost and the limited availability of fertilizer, grain production in SSA was limited to 124 kg person^{-1} compared with 237 kg person^{-1} in India, where fertilizer is more readily available, and 1136 kg person^{-1} in the United States, where fertilizer is both inexpensive and readily available (Palm et al., Chapter 5, this volume; FAO, 2004). In regions like North America, people consume near-double maintenance levels of both protein (114 g day^{-1} total) and calories (3700 kcal day^{-1} total), whereas many people within SSA have lower-than-needed protein and calories available for consumption.

The fact that N fertilizer is not used efficiently is in part responsible for these issues. On average the crop takes up only 20 to 50 percent of the N applied to soil for cereal crop production. Although N fertilizer use is low in many parts of the world, the NUE may be lower than in areas where consumption is higher. Low efficiency of N is typically caused by an insufficiency of other required nutrients (e.g., P, K, and secondary and micronutrients, Aulakh and Malhi, Chapter 13, this volume), which limits plant growth along with N. In rice production, NUEs of 30 percent or lower are typical in many regions, whereas efficiencies approaching 70 percent are not uncommon in areas of intensive maize production (Dobermann and Cassman, Chapter 19, this volume). Even in these high-efficiency regions, losses of N occur, exacerbating water-quality problems both locally and downstream of crop production areas.

Agricultural Nitrogen Cycle

Fertilizer supplies about 50 percent of the total N required for global food production. In 1996 global fertilizer N consumption totaled 83 Tg N (Smil 1999), and consumption has increased little since then, for example, 84.1 Tg N in 2002 (FAO 2004). Therefore, Smil's estimates of the global N flows are probably still appropriate and are used here. The other annual inputs into crop production—biological N-fixation (~33 Tg; 25–41 Tg), recycling of N from crop residues (~16 Tg; 12–20 Tg) and animal manures (~18 Tg; 12–22 Tg) (Figure 1.1), atmospheric deposition, and irrigation water (not shown in Figure 1.1)—provide an additional ~24 Tg (21–27 Tg) (Smil 1999). Of the ~170 Tg N added, about half is removed from the field as harvested crop (~85 Tg). The remainder of the N is incorporated into soil organic matter or is lost to other parts of the environment for which global estimates of individual loss vectors are highly uncertain. Leaching, runoff, and erosion account for ~37 Tg of the annual N losses; ammonia volatilization from soil and vegetation contributes ~21 Tg yr^{-1}. Denitrification losses as gaseous dinitrogen (N_2) amount to ~14 Tg yr^{-1}, and N_2O and NO from nitrification/denitrification contribute another ~8 Tg N to the total loss (Smil 1999; Balasubramanian et al., Chapter 2; Peoples et al., Chapter 4; Goulding, Chapter 15; Boyer et al., Chapter 16, this volume). Van der Hoek (1998) also estimated that more than 60 percent of the annual N input into food production was not converted into usable product. This *surplus N,* defined as the difference between input and output, is either lost to the environment or accumulates in the soil. Agricultural soils in the United States (and probably most of those in Western Europe) are considered to be at near steady state for soil accumulation of N; thus, all inputs not removed from the field in crops are likely to be lost to the atmosphere or aquatic systems (Howarth et al. 2002).

The relative inefficiencies of animal protein production exacerbate the inefficiencies of N utilization. Larger N losses from global food production are likely in the future as the human population and the demand for animal protein increase (Galloway et al. 2002a). The increase in consumption of animal products worldwide, except for regions within SSA, has been accompanied by an intensification of animal products in some regions, particularly North America. Because of the centralization of livestock production in regions that produce relatively little animal feed, the areas of crop production

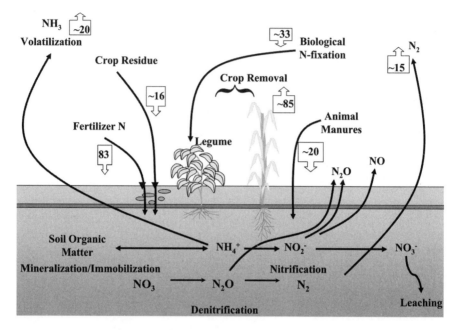

Figure 1.1. A simplified view of the nitrogen (N) cycle in crop production. Estimated global N flows (inputs and losses, Tg N yr^{-1}) are taken from Smil (1999).

located close to the intensive animal-production systems are not adequate to carry the load of animal waste input. As a result, the remainder of the N is stored in lagoons or solid piles (Smil 1999) or distributed elsewhere, partly through NH_3 volatilization, surface runoff, leaching, and wind erosion. Most of the volatilized NH_3 is deposited near the feedlot, but significant amounts can be converted to aerosols and transported 1000 km or farther. Much of the remaining "unused" N eventually finds its way into ground and surface waters. These losses can contribute to environmental and human health problems (Peoples et al., Chapter 4, this volume).

Environmental and Human Health Impacts

One of the most important impacts of N on the environment is that on water quality. Because N is frequently the nutrient most limiting biological productivity in estuaries (Vitousek et al. 1997), inputs of soil and fertilizer N from agricultural land can be a major contributor to N-induced eutrophication. The excessive growth of algae and macrophytes, the resulting oxygen depletion, and the production of a range of substances toxic to fish, cattle, and humans are now major pollution problems worldwide (Howarth et al. 1996). In contrast, low levels of N in soil can be a causative factor in soil erosion, which is a major contributor to land degradation. An insufficient amount of plant-available N can limit plant growth, resulting in reduced canopy interception of rainfall and less soil-binding by plant roots, both of which result in increased soil loss and can have major impacts on water quality through sedimentation and the release of N and P, causing excessive growth of aquatic nuisance plants.

According to Townsend et al. (2003), increases in reactive N in the environment have some clear and direct consequences for human health; air pollutants, primarily nitrogen oxides (NO_x) and dietary nitrate, have been issues of concern. In the case of dietary nitrate, much confusion and controversy remain (McKnight et al. 1999; Peoples et al., Chapter 4, this volume).

Almost 60 years ago, high nitrate (which can be reduced to nitrite in the intestine) concentrations in drinking water drawn from local wells (Comly 1945) were implicated in the incidence of infantile methemoglobinemia ("blue baby syndrome"). In recent years this view has been challenged, and strong evidence now exists that endogenous nitric oxide/nitrite production, triggered by intestinal infection rather than exogenous dietary nitrate intake, is responsible (McKnight et al. 1999; L'hirondel and L'hirondel 2002). This condition now appears to be rare in the developed world, where nitrate levels in drinking water are higher than they previously were and for the most part are increasing; in less-developed countries, ingestion of contaminated water, and its associated gastroenteritis, appears to be a more likely cause of methemoglobinemia (Leifert and Golden 2000).

The changing situation with regard to dietary nitrate and gastrointestinal cancer is equally interesting. Early thinking called for restrictions on nitrate levels in food because of the formation of carcinogenic nitrosamines by nitrosation of amines in the gastrointestinal tract (McKnight et al. 1999); however, not only is the incidence of gastric and intestinal cancers reduced in groups who consume vegetables high in nitrate (Corella et al. 1996), but there is also a worldwide decline in the incidence of gastric cancer (Correa and Chen 1994) at the same time the nitrate content and intake of green vegetables are increasing (McKnight et al. 1999). Epidemiologic studies point toward a possible protective effect of nitrate (L'hirondel and L'hirondel 2002). These studies suggest that dietary nitrate, which determines the production of reactive nitrogen oxide species in the stomach, is an effective host defense against gastrointestinal pathogens and can have beneficial effects against cancer and cardiovascular diseases.

The nitrate–human health issues remain controversial, and a thorough reevaluation is timely. This area is an important one for further work, given that nitrate levels in groundwater in Europe are sometimes larger than the currently recommended safe levels.

Prospects for Increasing Nitrogen Use Efficiency

As pointed out in several chapters of this volume, fertilizer N has a low efficiency of use in agriculture (10–50 percent for crops grown in farmers' fields; Balasubramanian et al.,

Chapter 2, this volume). One of the main causes of low efficiency is the large loss of N by leaching, runoff, ammonia volatilization, or denitrification (Raun and Johnson 1999), with resulting contamination of water bodies and the atmosphere. With the limitation on arable land area and the need to minimize the pollution of waters and the atmosphere with reactive N derived from N fertilizer, the only way to continue to feed the increasing population is to increase the efficiency of use of fertilizer N (Cassman et al. 2002).

It is important to know the forms and pathways of N loss and the factors controlling them so that procedures can be developed to minimize the loss and increase the NUE. Investigations have shown that the predominant loss process and the amounts lost are influenced by ecosystem type, soil characteristics, cropping and fertilizer practices, and prevailing weather conditions. As a consequence, losses can vary considerably over small distances within a field because of soil variability, from region to region because of differing cropping practices, and with time over a growing season because of climate. In Europe, where nitrate forms of fertilizer dominate, nitrate leaching and denitrification are the main loss pathways; in the rest of the world, where urea is the main fertilizer used, ammonia volatilization tends to be more important (Goulding, Chapter 15, this volume).

Volatilization of added N as ammonia from fertilized grassland (13 percent of added N), upland crops (18 percent), and fertilized rice (20 percent) in developing countries exceeds that lost in developed countries (6, 8, and 3 percent, respectively; IFA/FAO 2001). The largest losses overall and the lowest NUEs, however, tend to occur in the developed world (Goulding, Chapter 15; Dobermann and Cassman, Chapter 19, this volume). The low efficiency in the developed world occurs because farmers often apply excess N as insurance against low yields. The relatively low cost of fertilizer N compared with the value of the crop product lost in the developed world has led to its misuse and overapplication. The same does not usually hold true in the developing world, where access to fertilizer is limited (Hubbell 1995).

Many approaches have been suggested for increasing fertilizer NUE, including the optimal use of fertilizer form, the rate and method of application, matching N supply with crop demand, optimizing split application schemes, supplying fertilizer in the irrigation water, switching from urea to calcium ammonium nitrate to limit ammonia loss, minimizing application in the wet season to reduce leaching, applying fertilizer to the plant rather than to the soil, changing the fertilizer type to suit the conditions, and using slow-release fertilizers (Balasubramanian et al., Chapter 2, this volume). The genetic variation in both acquisition and internal-use efficiencies indicates potential for further increases in NUE through plant selection (Giller et al., Chapter 3, this volume).

In addition, agronomic practices that improve early crop growth, reduce competition for N uptake by weeds, reduce pest incidence, and improve irrigation and drainage will increase the NUE. Dobermann and Cassman (Chapter 19, this volume) provide an example of how such external factors, in addition to N management, can increase the

NUE. The factors involved in increasing this efficiency in corn production in United States from 42 to 57 kg grain kg N^{-1} were (1) greater stress tolerance of modern maize hybrids; (2) improved management (conservation tillage, better seed quality, higher plant densities, weed and pest control, balanced fertilization with other nutrients, irrigation); and (3) improved matching of the amount and timing of applied N to the indigenous supply and crop demand.

Lack of adequate rainfall for crop growth in semi-arid areas limits the extent to which crops can respond to fertilizer N, resulting in poor NUE. McCown et al. (1991) showed the benefit of linking fertilizer application to precipitation by using crop simulation modeling coupled with historical climate data in the Machakos district in semi-arid Kenya.

As pointed out by Dobermann and Cassman (Chapter 19, this volume; Figures 19.3 and 19.4), increased NUE has been achieved at the national scale, but current efficiencies on cereal cropping farms (20–50 percent; Cassman et al. 2002) are well below those reported in small-scale research plots (60–90 percent, Balasubramanian et al., Chapter 2, this volume). This difference is often explained by the better management of research plots with regard to water supply, weed and pest management, and balanced nutrition. Improving farm-scale management toward matching that on research plots would increase NUE and enhance environmental quality. We conclude that the best prospects for increased NUE lie with improved management of soil, water, crop, and fertilizer.

Contributors to the Food Production Chain

Primary agriculture is part of the food production chain in which six major contributors participate and influence each other (Figure 1.2, left side). When societies shift progressively from an agricultural to an industrialized to a service-providing society, the role and value (in monetary terms) of primary agriculture become smaller and the roles of suppliers (e.g., of N fertilizer, seeds), the processing industry, wholesale dealers, retailers, and consumers become larger. At the same time, the influence of contributors outside the production chain increases.

The food production chain of any country or region does not exist in isolation from other parts of the socioeconomic system. For example, government policies have a great influence on the effectiveness of local and regional infrastructure, on which primary agricultural production is heavily dependent (Figure 1.2; Palm et al., Chapter 5, this volume) for the delivery of inputs to the farm and the transport of products from the farm to local, national, or international markets.

Contributors outside the production chain often focus on one contributor within the production chain; but as the influence of suppliers, the processing industry, wholesalers, retailers, and consumers on the production process increases at the expense of the primary producers, contributors outside the production chain may also change their pri-

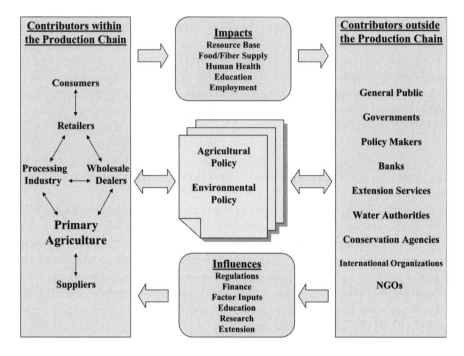

Figure 1.2. Interaction of contributors within and outside the production chain through impacts from the production chain and influences on it. The authors acknowledge the input of Dr. Jorgen Olesen in developing this figure.

mary focus. For example, water authorities and environmental nongovernmental organizations (NGOs) focus increasingly on the farming and processing industries with regard to the impact of nitrate on surface waters and on groundwater used for domestic water consumption (Peoples et al., Chapter 4, this volume). Much of this influence is, however, indirect through agricultural and environmental policies and extension (Figure 1.2). This process is iterative because the issues (e.g., pollution) that impact, for example, the ecosystem services provided by agriculture, such as food, air, and water, biodiversity, and landscape variability, then feed back through governmental policy decisions to influence agriculture. Thus the interplay of contributors, both within and outside the food production chain, requires different balance and interpretation among production, environmental, economic, and social functions in different regions (Palm et al., Chapter 5, this volume).

In the context of N fertilizer, agricultural and environmental policies have major effects in determining use in a given country and the effect that N fertilizer is likely to have on the production of food and on the environment. The impacts of N fertilizer use determine whether priority must be given to increasing fertilizer use, as in SSA, to increase food production and increase rural livelihoods (Vanlauwe et al., Chapter 8, this volume) or whether environmental and perceived health issues dictate the agenda and lead to a reduction in N fertilizer use. In both cases, more efficient use of fertilizer N is desirable (Dobermann and Cassman, Chapter 19, this volume).

Four basic elements and many contributors, inside and outside the production chain, are potentially involved in developing and implementing policies and strategies to improve fertilizer NUE and should be addressed coherently:

1. The policy instrument (regulation or stimulation of activity?)
2. The technical component (what action, what measure?)
3. The addressee (against whom is action taken or to whom are measures addressed?)
4. The spatial dimension of the policy/strategy (which area?)

For example, a tax on N fertilizer could be implemented with the supplier or primary producer; constraints on the production process could be introduced at the retailer stage, and the retailer then imposes constraints and targets on the processing industry, the wholesale dealers, and the primary producer. Likewise, incentives to increase N fertilizer use can be provided by reducing the financial cost of N fertilizer through suppliers or by increasing the value of the crop by providing price support through retailers and wholesale dealers.

An important feature emerges from this brief consideration. Any policies relating to N fertilizer use should be formulated jointly by the contributors both within and outside the production chain with a view to ensuring feasibility and optimizing effectiveness. This is because such policies can have direct and indirect (through contributors outside the production chain) impacts.

Who Pays for Protecting the Environment?

Too little or too much N fertilizer can contribute to human health and environmental problems. These problems come at high economic costs, are complex, and are not amenable to single solutions. The costs and benefits of environmental quality are difficult to determine (Moomaw 2002), and different views exist as to how these costs should be met (Palm et al., Chapter 5, this volume).

The issues of underuse and overuse of N fertilizer can be traced to three types of malnutrition that impact approximately two thirds of the global human population (~4.0 billion persons): (1) Deficiencies in calories and protein affect ~0.8 billion persons (FAO, 2004), (2) another ~2 billion persons have adequate caloric intake but suffer from vitamin and mineral deficiencies, and (3) the remaining ~1.2 billion persons have an unbalanced diet through consuming excess protein and calories and are overweight. The first two types of malnutrition are problems mainly of the developing world, whereas the third type is an issue of the developed world. Both deficiencies and overconsumption contribute to health problems that come at high economic and social costs (Gardner and Halweil 2000). Ironically, the problems of dietary deficiencies and inefficient use of fertilizer N contribute to human health problems, environmental degradation, and thus societal problems in similar ways. An inadequate nutrient supply promotes soil degradation through loss of soil organic matter, low biomass productivity, and increased soil erosion. Increasing fertilizer use in such situations (e.g., in SSA) promotes increased "land use efficiency" (Fixen and West 2002) and serves to increase food production and alleviate environmental problems. In the case of overproduction, increasing NUE contributes to decreasing nitrate loading of ground and surface water supplies. All three types of malnutrition are important human health issues.

Who pays for the costs associated with human health and environmental problems that are related to either too little or too much N fertilizer? Agriculture is one of the greatest users of our natural resources, including land, soil, water, and forests; and diverse interest groups are concerned with the management of these resources. Those mainly concerned are agricultural producers, conservationists, and people interested in their future and that of their descendants (Alex and Steinacker 1998). Farmers value water and soil resources because of the increasing costs of irrigation water and decreased productivity because of acidification, salinization, and erosion. Conservationists value the aesthetic and social benefits of natural resources and the environment, the social value of which has increased dramatically in recent years because deterioration of the environment has became more evident; and increased incomes, education, and leisure time have allowed a greater appreciation of natural landscapes and clean air and water. People who consider the future have concerns for the effect of agricultural activities on global warming, the ozone layer, the safety of our drinking water, and future food supplies (e.g., Alex and Steinacker 1998).

The question, then, is how the costs of conserving our natural resources and the environment should be apportioned among the interested parties. Governments (acting on behalf of the people) have a role in ensuring long-term production and the supply of adequate food supplies, developing and maintaining sustainable production systems, and protecting the environment. Conservationists have interests in preserving natural resources and the environment, and farmers need to increase production on a sustainable basis, maximize profits on investments, and conserve the natural resource base for future production. Governments in general have placed a high priority on protection of the environment, but this has not always been translated into action and financing by individual countries. In centrally planned economies, many environmental problems have not been addressed because the major focus has been on development (Alex and Steinacker 1998).

Nitrogen fertilizer also can impact more than one part of an ecosystem at the same time: for example, air quality as a result of dust from wind erosion; water erosion of soil because of a lack of ground cover and siltation of surface water supplies (undersupply), NO_x emission, and O_3 generation and nitrate leaching and runoff (over supply); and human health because of malnutrition (both undersupply and oversupply). Governmental policies are typically directed at one problem at a time rather than considering them in an integrated approach to human nutritional and environmental needs (Moomaw 2002).

So who should pay for the real costs of too little N fertilizer for food, feed, and fiber production and too much N fertilizer for environmental quality and human health? Should it be producers, consumers, governments, or a combination of all three? The answer is likely to differ with the situation, but whether the costs are hidden and paid through taxation or paid for by increased food costs, facing the issue directly may be the least expensive alternative over the long term. This issue has yet to be resolved and, given the complexities of the social, economic, environmental, and political dimensions involved, one that is far from easy. The International Nitrogen Initiative could usefully provide further insight into this in its future deliberations.

Literature Cited

Alex, G., and G. Steinacker. 1998. Investment in natural resources management research: experience and issues. Pp. 249–271 in *Investment strategies for agriculture and natural resources*, edited by G. J. Persley. Wallingford, UK: CABI Publishing.

Boyer, E. W., and R. W. Howarth. 2002. *The nitrogen cycle at regional to global scales.* Dordrecht, The Netherlands: Kluwer Academic Publishers.

Brown, L. R. 1999. Feeding nine billion. Pp. 115–132 in *State of the world 1999: A Worldwatch Institute report on progress toward a sustainable society*, edited by L.R. Brown, C. Flavin, and H. French, et al. New York: W.W. Norton & Company.

Cassman, K. G., A. Dobermann, and D. Walters. 2002. Agroecosystems, nitrogen-use efficiency, and nitrogen management. *Ambio* 31:132–140.

Comly, H. H. 1945. Cyanosis in infants caused by nitrates in well water. *Journal of the American Medical Association* 129:112–116.

Corella, D., P. Cortina, M. Guillen, and J. I. Gonzalez. 1996. Dietary habits and geographic variation in stomach cancer mortality in Spain. *European Journal of Cancer Prevention.* 5:249–257.

Correa, P., and V. W. Chen. 1994. Gastric cancer. *Cancer Surveys* 19:55–76.

Cowling, E. B., J. N. Galloway, C. S. Furiness, M. C. Barber, T. Bresser, K. Cassman, J. W. Erisman, R. Haeuber, R. W. Howarth, J. Melillo, W. Moomaw, A. Mosier, K. Sanders, S. Seitzinger, S. Smeudlers, R. Socolow, D. Walters, F. West, and Z. Zhu. 2001. Optimizing nitrogen management in food and energy production and environmental protection: Summary statement from the Second International Nitrogen Conference, October 14–18, 2001. Washington, D.C.: Ecological Society of America.

FAO (United Nations Food and Agricultural Organization). 2004. *FAO agricultural data bases are obtainable on the World Wide Web*: http://www.fao.org.

Fixen, P. E., and F. B. West. 2002. Nitrogen fertilizers: Meeting contemporary challenges. *Ambio* 31:169–176.

Galloway, J. N., E. B. Cowling, S. P. Seitzinger, and R. H. Socolow. 2002a. Reactive nitrogen: Too much of a good thing? *Ambio* 31:60–71.

Galloway, J. N., E. B. Cowling, J. W. Erisman, J. Wisniewski, and C. Jordan. 2002b. Optimizing nitrogen management in food and energy production and environmental protection. Pp.1–9 in *Proceedings of the 2nd International Nitrogen Conference on Science and Policy.* Potomac, Md, 14–18 Oct 2001. Exton, Penn.: A. A. Balkema Publishers.

Gardner, G., and B. Halweil. 2000. Nourishing the underfed and overfed. Pp. 59–78 in *State of the world 2000: A Worldwatch Institute report on progress toward a sustainable society,* edited by L. R. Brown, C. Flavin, H. French, et al. New York: W.W. Norton & Company.

Greenland, D. J. 1988. Soil organic matter in relation to crop nutrition and management. Pp. 85–89 in *Proceedings of the International Conference on the Management and Fertilization of Upland Soils in the Tropics and Subtropics.* Nanjing: Chinese Academy of Sciences.

Howarth, R. W., E. W. Boyer, W. J. Pabich, and J. N. Galloway. 2002. Nitrogen use in the United States from 1961–2000 and potential future trends. *Ambio* 31:88–96.

Howarth R.W., G. Billen, D. Swaney, A. Townsend, N. Jaworski, J. K. Lajtha, J. A. Downing, R. Elmgren, N. Caraco, T. Jordan, F. Berendse, J. Freney, V. Kudeyarov, P. Murdoch, and Z. L. Zhu. 1996. Regional nitrogen budgets and riverine N & P fluxes for the drainages to the North Atlantic Ocean: Natural and human influences. *Biogeochemistry* 35:75–139.

Hubbell, D. H. 1995. Extension of symbiotic biological nitrogen fixation technology in developing countries. *Fertilizer Research* 42:231–239.

IFA/FAO. 2001. *Global estimates of gaseous emissions of NH_3, NO and N_2O from agricultural land.* Rome: FAO.

Leifert, C., and M. H. Golden. 2000. A re-evaluation of the beneficial and other effects of dietary nitrate. *Proceedings of the International Fertiliser Society Proceedings* No. 456, York, UK, 22 pp.

L'hirondel, J., and J. L. L'hirondel. 2002. *Nitrate and man.* Wallingford, UK: CABI Publishing.

McCown, R. L., B. M. Wafula, L. Mohammed, J. G. Ryan, and J. N. G. Hargreaves. 1991. Assessing the value of seasonal rainfall predictor to agronomic decisions: The case of response farming in Kenya. Pp. 383–409 in *Climatic risk in crop production: Models and management for the semi-arid tropics and subtropics,* edited by R. C. Muchow, and J. A. Bellamy. Wallingford. UK: CAB International.

McKnight, G. M., C. W. Duncan, C. Leifert, and M. H. Golden. 1999. Dietary nitrate in man: Friend or foe? *British Journal of Nutrition* 81:349–358.

Moomaw, W. R. 2002. Energy, industry and nitrogen: Strategies for decreasing reactive nitrogen emissions. *Ambio* 31:184–189.

Mosier, A. R., M. A. Bleken, P. Chaiwanakupt, E. C. Ellis, J. R. Freney, R.B. Howarth, P. A. Matson, K. Minami, R. Naylor, K. N. Weeks, and Z. L. Zhu. 2002. Policy implications of human-accelerated nitrogen cycling. Pp. 477–516 in *The nitrogen cycle at regional to global scales,* edited by E. W. Boyer and R. W. Howarth. Dordrecht, The Netherlands: Kluwer Academic Publishers.

Raun, W. R., and G. V. Johnson. 1999. Improving nitrogen use efficiency for cereal production. *Agronomy Journal* 91:357–363.

Smil, V. 1999. Nitrogen in crop production: An account of global flows. *Global Biogeochemical Cycles* 13:647–662.

Smil, V. 2002. Nitrogen and food production: Proteins for human diets. *Ambio* 31:126–131.

Syers, J. K. 1997. Managing soils for long-term productivity. *Philosophical Transactions of the Royal Society of London (B)* 352:1011–1020.

Townsend, A. R., R. W. Howarth, F. A. Bazzaz, M. S. Booth, C. C. Cleveland, S. K. Collinge, A. P. Dobson, P. R. Epstein, E. A. Holland, D. R. Keeney, M. A. Mallin, C. A. Rogers, P. Wayne, and A. H. Wolfe. 2003. Human health effects of a changing global nitrogen cycle. *Front Ecology Environment* 1:240–246.

Van der Hoek, K. W. 1998. Nitrogen efficiency in global animal production. *Environmental Pollution* 102:127–132.

Vitousek, P. M., J. Aber, R. W. Howarth, G. E.Likens, P. A. Matson, D. W. Schindler, W. H. Schlesinger, and D. G. Tilman. 1997. Human alteration of the global nitrogen cycle: Causes and consequences. *Issues in Ecology* 1:1–15.

PART II
Crosscutting Issues

2

Crop, Environmental, and Management Factors Affecting Nitrogen Use Efficiency

Vethaiya Balasubramanian, Bruno Alves, Milkha Aulakh,
Mateete Bekunda, Zucong Cai, Laurie Drinkwater,
Daniel Mugendi, Chris van Kessel, and Oene Oenema

Nitrogen (N) is a key input to food production. The availability of relatively inexpensive N fertilizers from the 20th century onward has contributed greatly to increased food production, although not equally on all continents (Smil 2001). Currently about 40 percent of the human population rely on N fertilizer for food production. About 56 percent of the N fertilizer is used for producing rice, maize, and wheat (IFA 2002). These cereals and other crops use an average of 50 percent or less of the applied N for producing aboveground biomass (Krupnik et al., Chapter 14, this volume). The other 50 percent is mostly dissipated in the wider environment, causing a number of environmental and ecologic side effects (Galloway and Cowling 2002). These N losses are an economic loss to farmers, especially for smallholders in Africa, where fertilizer costs represent a large fraction of the total costs and where increases in food production are urgently needed (Sanchez and Jama 2002). Clearly, significant improvements must be made in N use efficiency (NUE) to produce enough food to feed the growing population and avoid large-scale degradation of ecosystems caused by excess N (Tilman et al. 2001).

This chapter deals with fertilizer NUE and factors controlling it in a number of major crop production systems. In field studies, four agronomic indices are commonly used to measure NUE: partial factor productivity (PFP_N), agronomic efficiency (AE_N), apparent recovery efficiency (RE_N), and physiologic efficiency (PE_N) as defined in the Appendix. For this chapter, we selected RE_N as an indicator of fertilizer NUE of crops and cropping systems but acknowledge other important sources of N must be considered in constructing a complete N budget for agriculture. Our purpose is to identify the

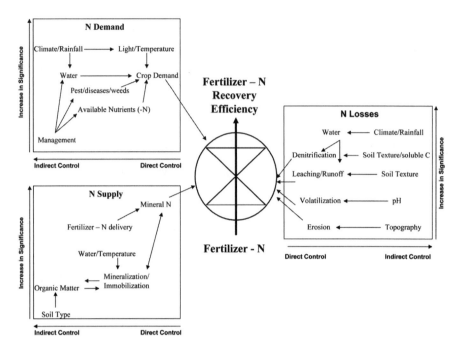

Figure 2.1. Conceptual model depicting the three main control boxes (i.e., nitrogen demand, supply, and losses) and their major processes and variables regulating fertilizer N use efficiency (NUE). The symbol in the center of the figure represents the "control center," which influences the flow of fertilizer N into the crop and therefore the apparent recovery efficiency of applied N (RE_N). The horizontal listing and their distance from the control center of the processes and variables within each box reflect their direct or indirect effect on RE_N. The vertical location of processes and variables within each box reflects their level of significance on RE_N. For further explanations, see text.

major factors limiting RE_N under field conditions and to identify opportunities for improving average RE_N values obtained under on-farm conditions.

Conceptual Model of Nitrogen Use Efficiency

Figure 2.1 presents as a conceptual model the key processes and variables that control the RE_N. Fertilizer RE_N by crops is driven by three main sets of controls: (1) crop N demand, (2) N supply, and (3) N losses. Each set of controls comprises several processes and variables. Some processes can be managed in a field (e.g., delivery of nutrients, disease control), but other variables cannot be controlled (temperature, rainfall, or soil texture).

The processes and variables that control the uptake of N by crops (and thus the RE_N as the control center in Figure 2.1) can exert a direct or an indirect effect on RE_N, and they can also be placed in an order of increasing significance. Hence, the processes and variables, which have a direct effect on RE_N and are placed at a high level of significance, will exert a major control on RE_N. In contrast, processes and variables operating at an indirect level and placed at a low level of significance will have less effect on RE_N.

Foremost, the demand for N drives the RE_N by a crop. Crop yield is highly correlated with total N uptake (Dobermann and Cassman, Chapter 19, this volume). Crop N demand is directly related to certain fundamental processes, associated with crop growth, that is, light (energy) and temperature (Loomis and Connor 1992). The availability of water and other nutrients (P, K, Mg, S) increases crop demand for N and RE_N (Smith and Whitfield 1990). The RE_N will further increase when insect pests, diseases, and weeds are eliminated.

Supply of N in soil originates from the application of N fertilizer or from net mineralization of soil organic matter (SOM) or crop residues. The RE_N depends partly on how much mineral N originated from current fertilizer application versus net mineralization of SOM or unused fertilizer N from previous applications (Figure 2.1). Of more significance in controlling RE_N, however, is the synchronization of N supply with crop demand for N. For example, split N application (Riley et al. 2003) could synchronize N supply with wheat crop demand for N, leading to higher RE_N.

By creating a strong sink for fertilizer N in the crop (i.e., removing all growth-limiting factors) and by providing an optimum delivery system of fertilizer N to the crop, a maximum RE_N value of 90 percent (assuming 10 percent of the acquired N remain in the roots) could theoretically be obtained. The theoretical maximum RE_N value, however, is never obtained because it is impossible to optimize all the factors that control crop N demand, N supply, and N losses. Fertilizer N can be lost through denitrification, leaching, runoff, volatilization, and soil erosion (see Figure 2.1).

Nitrogen Use Efficiency in Major Cropping Systems

It has been difficult to obtain RE_N values for many crops because of the scarcity of reliable data from farms or research trials. The RE_N values given in Table 2.1 for major cereals grown in intensive systems are likely to be more reliable than other crops, especially those in subsistence systems. Similarly, reliable RE_N values are not available even for major crops grown under rain-fed conditions. These problems with data reliability must be considered when interpreting the data in Table 2.1. Clearly, RE_N values for each crop vary considerably across regions because of differences in climate, soil type, and crop management.

Rice (Oryza sativa)

Globally, more than 90 percent of the rice is produced in Asia, using 93 percent of the total N fertilizers allocated for rice (FAO 2001). Irrigated rice receives much more fertilizer N than rain-fed rice.

About 75 percent of the global rice production comes from irrigated rice, which

Table 2.1. Mean recovery efficiency of nitrogen (RE_N) values of harvested crops under current farming practice and mean and maximum RE_N values obtained in research plots

Crops	Mean RE_N under current farming practice (%)	Mean RE_N in research plots (%)	Maximum RE_N of research plots (%)	Source of data
Rice				
Irrigated	31 (Asia)	49	88	Krupnik et al. Chapter 14, this volume
Rain-fed	20*	45*	55*	*Expert knowledge
Wheat				
Irrigated	33 (India)	45	96	Krupnik et al., Chapter 14, this volume
Rain-fed	17 (USA)	25	65*	Schlegel et al. 2003 *Expert knowledge
Maize				
Irrigated	37	42	88	Krupnik et al., Chapter 14, this volume
Rain-fed	30	40	65	Randall et al. (2003)
Vegetables	30*	50*	80	Singandhupe et al. 2003; *Expert knowledge
Root crops	10 (cassava) 40 (potato) 50 (sugar beet)	30 (cassava) 60 (potato) 70 (sugar beet)	40 (cassava) 70 (potato) 80 (sugar beet)	Hartemink et al. 2000; Neeteson 1989; Dilz 1988
Sugarcane	30	40	63	Basanta et al. 2003; Prasertsak et al. 2002
Cotton	35*	40	76	Rochester et al. 1997; *Expert knowledge
Coffee	40*	58*	80	Chaves, 2002; *Expert knowledge
Tea	10*	45*	55*	*Expert knowledge
Oil palm	50*	—	—	*Expert knowledge
Rubber	40*	—	—	*Expert knowledge
Non-grazed grassland	60	75	90	Dilz 1988; Whitehead 2000
Grazed grassland	5 (extensive) 15 (intensive)	15 (extensive) 30 (intensive)	30 (extensive) 50 (intensive)	*Expert knowledge
Organic cropping	—	—	—	—

* Based on N in harvested products (milk in intensively grazed grassland and meat in extensively grazed grassland).

Note: Values are based on literature data and expert knowledge. See explanation in text.

occupies 50 percent of the total rice growing area. National average N application rates vary from 56 kg ha^{-1} in Thailand to 180 kg ha^{-1} in China (FAO 2001). Generally, grain yield and RE_N are lower in the wet season than in the dry season as a result of adverse weather conditions and higher pest incidence (IRRI-CREMNET 2000). RE_N in irrigated rice can be increased by better synchronization of N application, with crop demand using a chlorophyll meter or leaf color chart (IRRI-CREMNET, 2000). Very little fertilizer is applied to rain-fed rice, and the RE_N in farmer's fields is estimated as 20 percent compared with 45 percent in research trials (see Table 2.1).

Wheat (Triticum aestivum)

Varietal differences in N use, the level of soil fertility, balanced use of nutrients, timing and rate of N application, tillage and early crop establishment, and weed/pest control influence crop growth, yield, and RE_N in irrigated wheat. During the grain filling period, wheat plants can lose N because of leaf senescence or leaching and volatilization of N from leaves. In addition, high temperature and low humidity during grain filling can reduce the remobilization efficiency of N to grain and post-anthesis N uptake from soil (Melaj et al. 2003). In Mexico, better fine-tuning of split N application with crop demand enhanced N uptake by wheat and reduced N losses (Riley et al. 2003). In India, real-time N management using the chlorophyll meter or leaf color chart increased wheat yield and NUE (Bijay-Singh et al. 2002).

All the above factors, plus topography/land form (concave or convex to trap rainfall), rainfall distribution/moisture availability, and weather conditions at grain filling, affect crop growth, grain yield, and NUE in rain-fed wheat. Therefore, varietal improvement and agronomic practices that promote deep rooting and tolerance to drought can increase wheat yield and RE_N. Minimum tillage with residues on the soil surface improves soil-moisture conservation and hence increases wheat yield and N uptake (Melaj et al. 2003). The average RE_N was 25 percent when urea ammonium nitrate solution (UAN, 28 percent N) was injected into the soil but was only 17 percent when UAN was broadcast (Schlegel et al. 2003).

Maize (Zea mays)

The rate, timing, and method of N application, soil type, tillage, weed and pest pressure, weather at grain filling, and crop rotation influence the growth, yield, and NUE of irrigated maize. A total of 170 kg N ha^{-1} applied in three splits was more efficient than a single preplant application of 500 kg N ha^{-1} in maize (Fernandez et al. 1998). Synchronizing split N application with crop demand enhanced RE_N in irrigated maize (Varvel et al. 1997). In the United States, the amount of maize grain produced per kilogram of applied N increased from 42 kg in 1980 to 57 kg in 2000 following the development of high-yielding hybrids, improved crop management, and crop need-based fertilizer N application. Little further increase in maize yield occurs following an increase in the N rate (mean $AE_N = 13$) because of the already high maize yield (average of 8.6 Mg ha^{-1} during 1999–2001) (Cassman et al. 2002).

In addition to factors affecting NUE in irrigated maize, the amount and distribution of rainfall in the growing season are critical for rain-fed maize. Yields range from 5.5 and 12.3 Mg ha^{-1} in the U.S. corn belt. Fall-applied N, especially without a nitrification inhibitor, is 10 to 15 percent less efficient than spring-applied N (Randall et al. 2003). Havlin et al. (1999) reported that N fertilizer placement enhances efficiency, with an RE_N of 42 percent for broadcast, 50 percent for surface band, and 68 percent for subsurface band application of urea ammonium nitrate to no-till maize in Kansas State. In West Africa, N-efficient maize varieties such as Oba super 2 are promoted to obtain relatively high yields and RE_N at both low and high N rates (Sanginga et al. 2003).

Vegetables

High rates of N fertilizers are applied to intensive vegetable systems; annual N application rates to vegetables in China exceed 1000 kg ha^{-1} (Zhang and Ma 2000). Large variations in RE_N values exist for different vegetables. Leafy vegetables with a shallow root system have lower RE_N values than other vegetables with deep root systems. Reported RE_N values for vegetables range from 10 percent to greater than 80 percent. Although the average RE_N values for intensive vegetable systems range from 30 to 60 percent, a high RE_N of 82.5 percent was reported for vegetables under drip irrigation (Singandhupe et al. 2003). RE_N values (crop uptake + soil N in 0–25 cm layer) ranged from 24 to 27 percent for rice–vegetable systems, in contrast to 37 to 55 percent for rice–grain legume systems in northern Philippines (Tripathi et al. 1997). Owing to excess N application and poor management, RE_N is low (10 percent) and soils often become saline in vegetable fields in China. Farmers periodically flood the vegetable plots to wash the salts below the plow layer. These management practices waste N fertilizer and water and pollute groundwater and the atmosphere.

Root Crops

The root/tuber crops discussed here include sweet potato (*Ipomomoea batatas*), cassava (*Manihot esculenta*), Irish potato (*Solanum tuberosum*), and sugar beet (*Beta vulgaris*). Both planting and harvesting of these crops require soil tillage, which may induce enhanced organic N mineralization and affect RE_N negatively. In the tropics the sweet potato and cassava tubers and tender tops/leaves (pot herb) are consumed, whereas sweet potato vines are used as fodder. Generally, the response of cassava and sweet potato to N fertilizer is poor and the RE_N is low: 10 to 30 percent (Hartemink et al. 2000).

Values of RE_N for potato range between 30 and 70 percent, with 10 to 20 percent of the acquired N in the tops (Neeteson 1989). Critical factors are N rates, nematode and virus problems, and drought. Sugar beet has an extensive root system that effectively scavenges N from soil. The RE_N is relatively high (60–80 percent), but more than half of the acquired N is in the tops and leaves that often are not harvested (Dilz 1988). N rates and drought affect RE_N values in sugar beet.

Cotton (Gossypium hirsutum)

Mean RE_N on cotton farms is about 30 percent, whereas the reported mean RE_N on research plots is 40 percent and the maximum RE_N is 76 percent (Rochester et al. 1997). Weed control, soil water availability, and N fertilizer management are critical in cotton. The correct timing and placement of N fertilizer improve the NUE by reducing ammonia volatilization and denitrification losses. The high N demand by cotton at 30 to 45 days after crop emergence must be met fully by timely N application to maximize yield and N uptake. Use of the petiole nitrate test (Havlin, Chapter 12, this volume) or the multi-spectral reflectance sensor (Sui et al. 1998) to diagnose the N status of cotton and the timely application of N fertilizer enhanced the yield and the NUE in cotton.

Sugarcane (Saccharum officinarum)

The planted sugarcane crop frequently responds poorly to N fertilizer application, probably because of mineralization of soil organic N and endophytic biological nitrogen fixation (BNF) (Boddey et al. 2003). Reported RE_N for planted cane varies from 0 to 40 percent. Response to applied N is higher for the ratoon crop than for planted cane, with a mean RE_N of 30 percent on farms and 40 percent for research plots (Basanta et al. 2003). An adequate water supply from rainfall or irrigation is the key to efficient use of N (Ingram and Hilton 1986). Application of ammonium sulphate and incorporation of urea minimizes N loss from sugarcane (Prasertsak et al. 2002) and increases RE_N to greater than 60 percent (Basanta et al. 2003).

Coffee (Coffea spp.)

Depending on coffee variety and the intensity of crop management, RE_N varies from 20 to 60 percent in farmers' fields and from 40 percent to 75 percent in research plots. Water availability, fertilizer N management, and SOM level are the major factors affecting N supply to the crop. In addition, high plant density improves yield and NUE in coffee. With a split N application (four to five times) to coffee in Brazil, the RE_N increased from 65 percent under the low traditional density of 2,000 plants ha^{-1} to 80 percent under a high density of 10,000 plants ha^{-1} (Chaves 2002).

Tea (Thea sinensis)

In tea plantations, high N rates are applied in split doses to induce new flushes for repeated harvests. For example, N fertilizer application rates to tea plantations are ranked as the highest in Japanese agriculture (Agriculture and Forestry Statistics Association 1991). Recovery of applied N by tea plants is generally low (10–45 percent), and it decreases with increasing N rates and increasing age of tea plantations. RE_N in tea is lower in summer than in spring and autumn. Acidic soil conditions inhibit ammonia volatilization and retard nitrification and denitrification processes. Thus, leaching and runoff are the major sources of N loss in tea plantations.

Oil Palm (Elaeis guineensis)

Oil palm plantations are intensively managed, and fertilizer costs account for more than half of the total production costs (Rankine and Fairhurst 1999); however, no fertilizer studies using control and fertilized plots of oil palm have been reported. From available N input/N output data, we estimate an RE_N of 50 percent for oil palm plantations.

Rubber (Hevea spp.)

Although rubber responds well to N fertilizer application, no proper studies using control and fertilized plots have been conducted. Our estimated RE_N for rubber is 40 percent.

Grasslands

Nongrazed, grassland systems comprise short-term leys and permanent grasslands with high N inputs, where the grass is cut and fed to housed ruminants, either as fresh grass or as silage. Grasslands are nonleaky systems and have a high RE_N (60–90 percent) when total N application does not exceed 300 kg ha^{-1} yr^1 and when the N fertilizer and animal manure are properly split applied two to five times per year at application rates of 30 to 150 kg N ha^{-1} (Whitehead 2000).

Approximately one third of the terrestrial biosphere area is grassland that provides forage for grazing animals. These grassland systems comprise (1) nonmanaged natural grasslands, (2) extensively managed grasslands used for meat production in temperate areas, and (3) intensively managed grasslands used mainly for milk production in temperate areas. The first two rely for their N supply mainly on BNF by leguminous species, whereas the latter depends on N fertilizer or BNF and animal manure (Whitehead 2000). Grazing animals exert large effects on N cycling and NUE through the localized return of 70 to 95 percent of the N in herbage via urine and dung depositions, which are prone to high N loss, and through grazing losses (about 20 percent) via trampling, smothering, and fouling of the grass (Jarvis et al. 1995). At the system level, using the N in animal products as the harvested N, the NUE is 15 to 30 percent for grasslands grazed by dairy cattle and 5 to 15 percent for grasslands grazed by beef cattle and sheep. When the N in the ingested grass is used as the harvested N, the NUE is 40 to 60 percent, without much difference between animal types.

Organic Cropping Systems

Nutrient management in organic systems is approached from an ecosystems perspective, which acknowledges the importance of plants, SOM, and soil organisms in regulating N availability and maintaining internal cycling capacity. The intention is to manage the full range of soil organic N reservoirs, particularly those with relatively long mean residence times that can be accessed by crops via microorganisms. As with conventional farming systems, studies of organic farms indicate that the balance between N additions and N harvested in the crop varies tremendously because of large variations in N additions (Watson et al. 2002). Generally, on commercial farms, grain systems operate with smaller N surpluses ($2-50$ kg N ha^{-1} yr^{-1}) compared with horticultural crops where surpluses of 90 to 400 kg N ha^{-1} yr^{-1} are reported (Watson et al. 2002). Long-term studies of organically managed cropping systems indicate that yields comparable to conventionally managed systems can be achieved while N losses are significantly reduced (Drinkwater et al. 1998). In these studies, a larger proportion of total N input was accounted for in organically managed compared with conventionally managed rotations. Understanding the underlying mechanisms that enable some organically managed cropping systems to achieve high yields while reducing N losses will contribute to improving the management of inorganic N fertilizers (Drinkwater, Chapter 6, this volume).

Crop, Environmental, and Management Effects

The consideration of RE_N of crops and cropping systems indicates that crop characteristics, environmental factors, and management affect the RE_N. The effect of crop characteristics on RE_N is greatly modified by environmental and management factors. Clearly, differences in RE_N values for similar crops or cropping systems across locations are due to differences in climate, soil type, and crop management. Within a location, annual or seasonal variations in RE_N are caused by annual and seasonal changes in climate and the inability of farm managers to predict and timely respond to such changes in weather conditions during the growing season.

Crop Effects

Crops and crop varieties differ in their abilty to acquire N from soil (N uptake efficiency), in producing economic biomass per unit of N acquired (PE_N), and in harvest index. These variations in crop capabilities lead to differences in the average RE_N values of crops and crop varieties by a factor of two (Table 2.1). Generally, perennial crops have a higher RE_N than annual crops. Among annual crops, cereals often have greater RE_N than root crops, which in turn have higher RE_N than leafy vegetables. In addition, genetically modified pest-resistant Bt (*Bacillus thuringiensis*) crops such as maize and cotton produced higher yield and net profit in the United States (Havlin, Chapter 12, this volume); such increases in yield will increase RE_N if more N is not applied to obtain higher yields; however, efficient N use is rarely a major consideration in the choice of crops to be grown (Kurtz et al. 1984).

Environmental Effects

Photosynthetic active radiation (PAR) is the major driving force for crop growth and crop N demand (Figure 2.1), but it does not contribute much to spatial and temporal variations in RE_N in temperate zones. In the tropics, however, systematic differences in RE_N have been found for rice grown in dry and wet seasons, and these have been ascribed to differences in radiation, flood control, and pest incidence. The other two factors that affect crop growth and RE_N are temperature and rainfall, and these are highly variable both in space and time. Overall, the relative importance of environmental factors affecting RE_N is in the order of rainfall > temperature > irradiance, although strong interactions exist among these factors and soil type.

Management Effects

Management is often called the fourth production factor, after land, labor, and capital. The importance and complexity of management have increased greatly during recent years. Variations in RE_N among farms in a similar environment with similar soil type are due to differences in management. Management aspects that specifically influence RE_N are crop rotations and cover crops, soil tillage, weed and pest control, irrigation and drainage, and integrated nutrient use, as further discussed in the following sections.

CROP ROTATIONS AND COVER CROPS

Differences in crop management through selection and care of seeds and seedlings, time of planting and harvesting, pest control, and intensification of crop rotations contribute to variations in RE_N. Crop rotations may have indirect effects on NUE by improving soil physical conditions and by the so-called crop sequence effect, which may involve a whole set of different factors (Kurtz et al. 1984) and by building up SOM (Sisti et al. 2004). Inclusion of cover crops in any rotation improves NUE by the ability of some cover crops to recover residual N leached below the root zone of cash crops (Olesen et al., Chapter 9, this volume). Organic residues from cover crops or manures positively interact with applied fertilizer N and increase RE_N (Vanlauwe et al. 2002).

SOIL TILLAGE

Conventional and conservation tillage are the two principal strategies for tillage. The effect of conservation tillage on crop yields and RE_N is highly conflicting because of differences in weather condition, soil type, method of crop establishment, management of surface residues, and the occurrence of soil pathogens and weeds (Camara et al. 2003). Because the amount of N fertilizer applied does not generally increase in these systems, the RE_N should be higher under zero tillage than under conventional tillage. In southern Brazil, after 7 years of zero tillage, organic matter in surface soil (0–10 cm) increased significantly and the N rate applied to maize for a yield goal of 7 Mg ha^{-1} decreased from 150 to 75 kg N ha^{-1} starting from the fifth year after the introduction of zero tillage, suggesting a strong improvement in RE_N (Boddey et al. 1997).

WEED AND PEST MANAGEMENT

Weed emergence time, weed density, and weed relative volume determine the extent of yield loss (Conley et al. 2003) and thus RE_N. Insect pests and diseases, if not controlled adequately, will reduce crop yields and RE_N. Integrated pest management (IPM) practices aim at reducing pest damage in a cost-effective, safe, environmentally sensitive, and sustainable manner. The effect of Bt-resistant crops on RE_N is still unknown.

IRRIGATION AND DRAINAGE

Irrigation is the second most important factor after high-yielding varieties that contributed to tripling of the yields of major cereals during the past three to four decades. Maize yields in the United States reached more than 16 Mg ha^{-1} in research plots and 10 Mg ha^{-1} in farmers' fields consistent with high RE_N through precise irrigation, fertilization, and crop management (Dobermann and Cassman, Chapter 19, this volume). Irrigated rice farmers in Asia must allow the floodwater to disappear before topdressing N fertilizers and then irrigate to move the applied N to the root zone to prevent ammonia volatilization. Farmers in China apply about 50 percent of the N fertilizer preplant, which leads to high N losses as a result of low plant uptake and possibly leaching (Cai et al. 2002; Buresh et al., Chapter 10, this volume). Wherever possible, farmers must avoid drainage immediately after N fertilization. In many irrigated areas, lack of proper drainage, overexploitation of groundwater, and use of poor-quality water for irrigation contribute to soil salinization, which reduces crop yield and RE_N.

INTEGRATED NUTRIENT MANAGEMENT

Integrated nutrient management (INM) denotes the optimum use of all available nutrient sources: SOM, crop residues, manures, BNF, and mineral fertilizers. INM is the key component of integrated soil fertility management (ISFM) in Africa (Vanlauwe et al., Chapter 8, this volume). In addition to other practices, ISFM advocates the combined application of organics and mineral fertilizers to maximize crop yields and RE_N. Generally, organic residues and manures positively interact with applied fertilizer N and increase its efficiency (Vanlauwe et al. 2002; Olesen et al., Chapter 9, this volume).

The synergistic interaction of N with P, K, S, and several micronutrients can lead to considerable improvements in yield and RE_N (Aulakh and Malhi, Chapter 13, this volume). In contrast, crop response to N is poor or even negative in P- and K-deficient soils, resulting in low RE_N. Unbalanced N-P_2O_5-K_2O ratios (e.g., 100-36-19 for China, 100-37-12 for India, 100-35-45 for the United States) often diminish plant utilization of applied N and thus reduce the RE_N (Norse 2003). The desirable N-P_2O_5-K_2O ratio is 100-50-25/50 for cereal crops (PPIC-India 2000).

The commonly used surface broadcasting of ammonium fertilizers entails enormous N losses from the system and reduces N supply to crops (Randall et al. 1985). Humphreys et al. (1992) noted that RE_N was 37 percent for broadcasting, 46 percent for banding, and 49 percent for deep point placement of urea super granules (USG) in direct-seeded rice in Australia.

Conclusions

Fertilizer N will continue to play a key role in food production in the near future. Therefore, appropriate farming methods and strategies are needed to use N fertilizers as efficiently as possible. In any system, variations in crop demand for N, N supply to the crop, and N losses determine the efficiency of applied fertilizer N (indicated by the recovery efficiency of applied fertilizer N, RE_N). Reliable RE_N data are needed for crops other than major cereals in irrigated systems and all crops in rain-fed systems to improve fertilizer use efficiency. Crop characteristics and environmental and management factors greatly influence the RE_N. A good understanding of these three factors and their interactions is a prerequisite to design successful strategies for improving RE_N. The relative importance of environmental factors affecting RE_N is in the order of rainfall $\geq$ temperature $\geq$ irradiation. There are, however, strong interactions between these abiotic factors and soil type; a significant part of the variations between fields, farms, and regions is therefore attributed to the interactions of weather conditions, soil type, and farm management. Here again, further research is needed to obtain hard data on the effect of management practices on RE_N.

Through improvements in nutrient and crop/farm management practices, more potential for improving RE_N is possible than through improvements in fertilizer technology. Improving farm management is not an easy task, however; it requires appropriate technologies and decision support tools and adequate training for their proper use. Farmers or farm managers prefer technologies and tools that are simple and easy to use, that require minimum additional labor and time, and that are cost-effective. Finally, achievement of a widespread increase in RE_N requires the active collaboration of farmers, extension personnel, researchers, and governments.

Literature Cited

Agriculture and Forestry Statistics Association. 1991. *Fertilizer handbook.* Tokyo: Ministry of Agriculture, Forestry and Fisheries, Fertilizer and Machine Division. (In Japanese.)

Basanta, M. V., D. Dourado-Neto, K. Reichardt, O. O. S. Bacchi, J .C. M. Oliveira, P. C. O. Trivelin, L. C. Timm, T. T. Tominaga, V. Correchel, F. A. M. Cassaro, L. F. Pires, and J. R. de Macedo. 2003. Management effects on nitrogen recovery in a sugarcane crop grown in Brazil. *Geoderma* 116:235–248.

Bijay-Singh, Yadvinder-Singh, J. K. Ladha, K. F. Bronson, V. Balasubramanian, J. Singh, and C. S. Khind. 2002. Chlorophyll meter and leaf color chart-based nitrogen management for rice and wheat in northwestern India. *Agronomy Journal* 94:821–829.

Boddey, R. M., J. C. M. Sá, B .J. R. Alves, and S. Urquiaga. 1997. The contribution of biological nitrogen fixation for sustainable agricultural systems in the tropics. *Soil Biology and Biochemistry* 29:787–799.

Boddey R. M., S. Urquiaga, B .J. R. Alves, and V. Reis. 2003. Endophytic nitrogen fixation in sugarcane: Present knowledge and future applications. *Plant and Soil* 252:139–149.

Cai, G. X., D. L. Chen, H. Ding, A. Pacholski, X. H. Fan, and Z. L. Zhu. 2002. Nitrogen losses from fertilizers applied to maize, wheat, and rice in the North China Plain. *Nutrient Cycling in Agroecosystems* 63:187–195.

Camara, K. M., W. A. Payne, and P. E. Rasmussen. 2003. Long-term effects of tillage, nitrogen, and rainfall on winter wheat yields in the Pacific Northwest. *Agronomy Journal* 95:828–835.

Cassman, K. G., A. Doberman, and D. T. Walters. 2002. Agroecosystems, nitrogen-use, efficiency, and nitrogen management. *Ambio* 31:132–140.

Chaves, J. C. D. 2002. *Manejo do solo—adubação e calagem, antes e após a implantação da lavoura cafeeira.* Circular No. 120. Londrina, Brazil: IAPAR.

Conley, S. P., L. K. Binning, C. M. Boerboom, and D. E. Stoltenberg. 2003. Parameters for predicting giant foxtail cohort effect on soybean yield loss. *Agronomy Journal* 95:1226–1232.

Dilz, K. 1988. Efficiency of uptake and utilization of fertilizer nitrogen by plants. Pp. 1–26 in *Nitrogen efficiency in agricultural soils,* edited by D. S. Jenkinson and K. A. Smith. London: Elsevier.

Drinkwater, L. E., P. Wagoner, and M. Sarrantonio. 1998. Legume-based cropping systems have reduced carbon and nitrogen losses. *Nature* 396:262–265.

FAO (Food and Agriculture Organization). 2001. *FAOSTAT Database Collections.* http://www.apps.fao.org. Rome, Italy: FAO.

Fernandez, J., J. Murilo, F. Moreno, F. Cabrera, and E. Fernandez-Boy. 1998. Reducing fertilization for maize in southwest Spain. *Communication in Soil Science and Plant Analysis* 29:2829–2840.

Galloway, J. N., and E. B. Cowling. 2002. Reactive nitrogen and the world: 20 years of change. *Ambio* 31:64–71.

Hartemink, A. E., M. Johnston, J. N. O. O'Sullivan, and S. Poloma. 2000. Nitrogen use efficiency of taro and sweet potato in the humid lowlands of Papua New Guinea. *Agriculture, Ecosystems and Environment* 79:271–280.

Havlin, J. L., J. D. Beaton, S. L. Tisdale, and W. L. Nelson. 1999. *Soil fertility and fertilizers*. Upper Saddle River, New Jersey: Prentice Hall.

Humphreys, E., P. M. Chalk, W. A. Muirhead, and R. J. G. White. 1992. Nitrogen fertilization of dry-seeded rice in south-east Australia. *Fertilizer Research* 31:221–234.

IFA (International Fertilizer Industry Association). 2002. *Fertilizer use by crops*. 5th ed. Rome, Italy: IFA, IFDC, IPI, PPI, FAO. http://www.fertilizer.org/ifa/statistics/crops/fubc5ed.pdf

Ingram, K. T., and H. W. Hilton. 1986. Nitrogen-potassium fertilization and soil moisture effects on growth and development of drip-irrigated sugarcane. *Crop Science* 26:1034–1039.

IRRI-CREMNET (International Rice Research Institute—Crop and Resource Management Network). 2000. *Progress Report for 1998 & 1999*. Los Baños, Philippines: IRRI.

Jarvis, S. C., D. Scholefield, and B. F. Pain. 1995. Nitrogen cycling in grazing systems. Pp. 381–419 in *Nitrogen fertilization in the environment*, edited by P. E. Bacon. New York: Marcel Dekker.

Kurtz, L. T., L. V. Boone, T. R. Peck, and R. G. Hoeft. 1984. Crop rotations for efficient nitrogen use. Pp. 295–306 in *Nitrogen in crop production*, edited by R. D. Hauck. Madison, Wisconsin: ASA/CSSA/SSSA.

Loomis, R. S., and D. J. Connor. 1992. *Crop ecology in productivity and management in agricultural systems*. Cambridge: Cambridge University Press.

Melaj, M. A., H. E. Echeverria, S C. Lopez, G. Studdert, F. Andrade, and N. O. Barbaro. 2003. Timing of N fertilization in wheat under conventional and no-tillage system. *Agronomy Journal* 95:1525–1531.

Neeteson, J. J. 1989. Evaluation of the performance of three advisory methods for nitrogen fertilization of sugar beet and potatoes. *Netherlands Journal Agricultural Science* 37:143–155.

Norse, D. 2003. Fertilizers and world food demand implications for environmental stresses. Paper presented at the IFA-FAO Agriculture Conference *Global food security and the role of sustainable fertilization*. 26–28 March 2003. Rome, Italy: FAO.

PPIC–India (Potash and Phosphate Institute of Canada–India Program). 2000. The challenging face of balanced fertilizer use, in *Fertilizer knowledge, no. 2*. New Delhi: PPIC–India.

Prasertsak, P., J. R. Freney, O. T. Denmead, P. G. Saffigna, B. G. Prove, and J. R. Reghenzani. 2002. Effect of fertilizer placement on nitrogen loss from sugarcane in tropical Queensland. *Nutrient Cycling in Agroecosystems* 62:229–239.

Randall, G. W., J. A. Vetsch, and J. R. Huffman. 2003. Corn production on a sub-surface drained mollisol as affected by time of nitrogen application and nitrapyrin. *Agronomy Journal* 95:1213–1219.

Randall, G. W., K. L. Wells, and J. J. Hanway. 1985. *Modern technology and use*. 3rd ed. Madison, Wisconsin: Soil Science Society of America.

Rankine, I., and T. H. Fairhurst. 1999. *Field handbook. Oil Palm Series*, Volumes 1, 2, & 3. Singapore: Potash & Phosphate Institute (PPI).

Riley, W. J., I. Ortiz-Monasterio, and P. A. Matson. 2003. Nitrogen leaching and soil nitrate, nitrite, and ammonium levels under irrigated wheat in Northern Mexico. *Nutrient Cycling in Agroecosystems* 61:223–236.

Rochester, I. J., G. A. Constable, and P. G. Saffigna. 1997. Retention of cotton stubble enhances N fertilizer recovery and lint yield of irrigated cotton. *Soil and Tillage Research* 41:75–86.

Sanchez, P. A., and B. A. Jama. 2002. Soil fertility replenishment takes off in East and Southern Africa, in *Integrated plant nutrient management in sub-Saharan Africa: From concept to practice,* edited by B. Vanlauwe, J. Diels, N. Sanginga, and R. Merckx. Wallingford, UK: CABI Publishing.

Sanginga, N., K. Dashiel, J. Diels, B. Vanlauwe, O. Lyasse, R. J. Carsky, S. Tarawali, B. Asafo-Adje, A. Menkir, S. Shulz, B. B. Singh, D. Chikoye, D. Keatinge, and R. Ortiz. 2003. Sustainable resource management coupled to resilient germplasm to provide new intensive cereal-grain legume-livestock systems in the dry savanna. *Agriculture, Ecosystems and Environment* 100:305–314.

Schlegel, A. J., K. C. Dhuyvetter, and J. L. Havlin. 2003. Placement of UAN for dryland winter wheat in the Central High Plains. *Agronomy Journal* 95:1532–1541.

Singandhupe, R. B., G. G. S. N. Rao, N. G. Patil, and P. S. Brahmanand. 2003. Fertigation studies and irrigation scheduling in drip irrigation system in tomato crop (*Lycopersicon esculentum* L.) *European Journal of Agronomy* 19:327–340.

Sisti, C. P. J., H. P. dos Santos, R. Kohhann, B. J. R. Alves, S. Urquiaga, and R. M. Boddey. 2004. Change in carbon and nitrogen stocks in soil under 13 years of conventional or zero tillage in southern Brazil. *Soil and Tillage Research* 76:39–58.

Smil, V. 2001. *Enriching the earth.* Cambridge, MA: The MIT Press.

Smith, C. J., and D. M. Whitfield. 1990. Nitrogen accumulation and redistribution of late applications of ^{15}N-labelled fertilizer by wheat. *Field Crops Research* 24:211–226.

Sui, R., J. B. Wilkerson, W. E. Hart, and D. D. Howard. 1998. Integration of neural networks with a spectral reflectance sensor to detect nitrogen deficiency in cotton. ASAE Paper no. 983104. ASAE. 2950 Niles Road, St. Joseph, Michigan 49085–9659.

Tilman, D., J. Fargione, B. Wolff, C. D'Antonio, A. Dobson, R. W. Howarth, D. Schindler, W. H. Schlesinger, D. Simberloff, and D. Swackhamer. 2001. Forecasting agriculturally driven global environmental change. *Science* 292:281–284.

Tripathi, B. P., J. K. Ladha, J. Timsina, and S. R. Pascua. 1997. Nitrogen dynamics and balance in intensified rainfed lowland rice-based cropping systems. *Soil Science Society of America Journal* 61:812–821.

Vanlauwe, B., J. Diels, K. Aihou, E. N. O. Iwuafor, O. Lyasse, N. Sanginga, and R. Merckx. 2002. Direct interactions between N fertilizer and organic matter: Evidence from trials with ^{15}N labelled fertilizer, Pp. 173-184 in *Integrated plant nutrient manageemnt in sub-Saharan Africa: From concept to practice,* edited by B. Vanlauwe, J. Diels, N. Sanginga, and R. Merckx. Wallingford, UK: CABI Publishing.

Varvel, G. E., J. S. Schepers, and D. D. Francis. 1997. Ability for in-season correction of nitrogen deficiency in corn using chlorophyll meter. *Soil Science Society of Ameirca Journal* 61:1233–1239.

Watson, C. A., H. Bengtsson, M. Ebbesvik, A. K. Loes, A. Myrbeck, E. Salomon, J. Schroder, and E. A. Stockdale. 2002. A review of farm-scale nutrient budgets for organic farms as a tool for management of soil fertility. *Soil Use and Management* 18:264–273.

Whitehead, D. C. 2000. *Nutrient elements in grassland: Soil–plant–animal relationships.* Wallingford, UK: CABI Publishing.

Zhang, F., and W. Ma. 2000. The relationship between fertilizer input level and nutrient use efficiency. *Soil and Environmental Science* 9:154–157 (In Chinese.)

3

Emerging Technologies to Increase the Efficiency of Use of Fertilizer Nitrogen

Ken E. Giller, Phil Chalk, Achim Dobermann, Larry Hammond, Patrick Heffer, Jagdish K. Ladha, Phibion Nyamudeza, Luc Maene, Henry Ssali, and John Freney

Major drivers of change that affect agriculture and demands for food are rising population densities, globalization and liberalization of trade, climate change, and environmental concerns. These factors also act as drivers for the development and adaptation of technologies for increasing the efficiency of use of fertilizer N (NUE). Agricultural practices have not developed at the same pace in all regions of the world, so technologies that are readily available in some countries may be regarded as emerging in other areas (Hubbell 1995).

In this chapter we consider technologies that may increase the NUE of fertilizer N in the future. These technologies can be divided into two main groups:

1. Those related to the choice of crop species and the genetic enhancement of the plant that essentially determine the N "demand" side.
2. Those concerned with the management options that determine the availability of soil and fertilizer N for plant uptake.

We describe innovative approaches that may result in the better use of existing knowledge and conclude by considering future prospects for improving the efficiency of fertilizer N use.

Efficient Plants

The NUE is a complex trait with many components, and a great degree of compensation takes place among the components. Therefore, crop selection is based mostly on aggregate traits (acquisition and internal efficiency) over a range of N rates. NUE measured for different crops or genotypes has three components:

1. Efficiency of acquisition or recovery of soil and fertilizer N = plant N uptake per unit N supply.
2. Internal efficiency (IE) with which N is used to produce biomass (IE biomass) = plant biomass/plant N content.
3. The IE with which N is used to produce grain (IE grain) = grain yield/plant N content.

Acquisition Efficiency

Depending on the crop, differences in N acquisition may result from variation in (1) the interception of N and the ability to absorb N from various soil depths (e.g., Tirol-Padre et al. 1996); (2) the efficiency of absorption and assimilation of ammonium and nitrate; and (3) root-induced changes in the rhizosphere affecting N mineralization, transformation, and transport (Kundu and Ladha 1997). Variability in the interception of N is related to rooting characteristics such as root length, branching, and distribution that allow the plant to explore a greater volume of soil. The rate of uptake at the root surface does not seem to limit N uptake and appears to offer less opportunities for genetic improvement. For example, short-term measurements of root N uptake capacity by rice (*Oryza sativa* L.) suggested daily rates of uptake of up to 10 kg N ha^{-1} day^{-1} (Peng and Cassman 1998), which exceed by a large margin the daily uptake requirements to satisfy biomass accumulation. Relatively little is known about the effects of root-induced changes in the rhizosphere on N transformations and whether such effects might be amenable to manipulation.

Efficiency of Internal Nitrogen Use

Generally, the curvilinear relationship between crop biomass production and tissue N concentration is a close inverse one, with little variation in this relationship between crop species within categories of C_3 and C_4 photosynthesis (Greenwood et al. 1990). Internal N use efficiency is tightly linked with the harvest index so that crop improvements in harvest index automatically result in improvement in internal efficiency. Although crop varieties within a species may display genetic differences in grain protein content that are consistently expressed across different levels of N supply, relatively little genetic variation is found in the efficiency with which acquired N is converted to grain yield within a crop species (Cassman et al. 2003). Therefore, the potential to improve internal N efficiency with regard to grain yield may be limited apart from selecting varieties for lower grain N concentration. For many crops this may not be a viable option because grain protein concentrations determine end-use quality (such as bread or

Durum wheat) and because many low-income consumers derive most of their protein intake from grain.

Potential for Genetic Enhancement of Nitrogen Use Efficiency

Although in the past plant breeders have concentrated on improving potential yields, there is new emphasis on a number of topics including the nutritional value of foods (protein content in grain, essential amino acids, other minerals, etc.), reducing post-harvest losses, making crops more tolerant of stresses (cold, drought, salt), or reducing reliance on pesticides. Crop improvement approaches that will increase yield stability and reduce yield losses contribute to increasing the efficiency with which fertilizer N is converted into economic products.

The genetic variation in both acquisition and internal-use efficiencies (e.g., harvest index) indicates potential for further increases in NUE through plant selection, particularly in crops that have received less attention from breeders. For example, the potential for breeding for NUE may be much greater in cereals such as t'ef (*Eragrostis tef*), a major staple food in the Horn of Africa, and in other food crops such as vegetables.

Systems with different production goals, such as organic agriculture, require the development of varieties with different characteristics. Many vegetables, such as onions (*Allium cepa*), have small root-length densities and thus poor soil exploration, presumably because they have been selected for production under conditions of nutrient surplus in heavily fertilized or manured soils. Natural variation for root traits is limited in the onion germplasm, although the old onion cultivars had a higher root-length density compared with modern ones (De Melo 2003). Interspecific crosses between *Allium cepa* and its relatives *A. roylei* and *A. fistulosum* (bunching onion) show great promise for increasing root depth, root branching, and root-length density and thus soil exploration, which should lead to greater NUE in the future. Increasingly, emphasis is on breeding maize and wheat for low N environments, such as those in sub-Saharan Africa, which is resulting in strong advances in NUE (Bänziger and Cooper 2001).

Functional genomics and marker-assisted selection offer great promise in accelerating the rate of advances in genetic improvement. Transgenic crops that prevent yield losses (e.g., BT-cotton) contribute substantially to the economic NUE.

Enhancement of Dinitrogen Fixation in Non-legumes

Research on the potential contribution from free-living N_2-fixing bacteria, heterotrophic N_2-fixation in the rhizosphere of cereals and non-legumes (often termed *associative N_2-fixation*), and by endophytic N_2-fixing bacteria within non-legumes remains controversial. Because contributions are difficult to measure, even under the most favorable environments (e.g., sugarcane [*Saccharum officinarum* L.] in tropical environments), it is likely that inputs are less than 20 kg N ha^{-1} yr^{-1} (Giller and Merckx 2003) and will not be amenable to manipulation.

Substantial interest has developed in the incorporation of the mechanism for N fixation into non-N$_2$-fixing plants since the early 1980s, although in fact relatively little research has been conducted. Two basic approaches have been used: (1) to incorporate the nitrogenase enzyme directly into the plant, with the chloroplast as a likely target organelle; and (2) to engineer or stimulate the plants to nodulate with N$_2$-fixation bacteria (Ladha and Reddy 2000). Despite claims to the contrary, the prospect of N$_2$-fixing cereal crops remains distant, particularly given the lack of research on this topic.

Efficient Management Practices

The efficiency of use of fertilizer N can be improved by modifying the form of N applied and by changing the way it is used on the farm.

Efficient Fertilizers

N fertilizers predominantly contain N in the form of ammonia, nitrate, or urea (Roy and Hammond, Chapter 17, this volume). Specialty products that are basically modifications of the previously mentioned products (e.g., granular, liquid, or suspended forms, controlled release compounds, or fertilizers containing urease and nitrification inhibitors or other essential nutrients) have been and continue to be developed by both the public sector and private sector groups. Given the chemistry of N, however, it seems unlikely that forms of N fertilizer based on compounds other than ammonia, nitrate, or urea will be adopted in the foreseeable future. It seems that development of products containing alternative forms of N has been curtailed in the public sector as a result of a lack of research funding, but we understand that research of this type protected by nondisclosure agreements is in progress in the private sector.

New technologies employing controlled-release fertilizers and nitrification inhibitors have the potential to reduce N loss markedly and to improve NUE (Shaviv 2000). In the development of controlled-release N fertilizers, the emphasis now is on synchronizing the release of N with the demand of the crop, and this has resulted in the intensive use of polymer-coated urea in Japanese rice fields (Shoji and Kanno 1995). The supply of N by a single application of controlled-release fertilizer is expected to satisfy plant requirements and yet maintain low concentrations of mineral N in the soil throughout the growing season. As a result, labor and application costs should be cheap, N loss should be minimized, NUE should increase, and yields should be improved.

Shoji et al. (2001) showed that the use of controlled-release fertilizer instead of urea in a potato (*Solanum tuberosum* L.) field markedly increased tuber yields and NUE from 17.3 to 58.4 percent. Increased NUE has also been obtained in rice (Fashola et al. 2002) and direct-seeded onions (Drost et al. 2002). The use of controlled-release fertilizer has almost doubled over the past 10 years, but it still accounts for only 0.15 percent of the total fertilizer N used (Trenkel 1997). The main reason for the limited use of controlled-release fertilizer is the high cost, which may be 3 to 10 times the cost of conventional fertilizer (Shaviv 2000).

Maintaining the N in the soil as ammonium would prevent the loss of N by both nitrification and denitrification. One method of doing this is to add a nitrification inhibitor with the fertilizer. Acetylene is a potent inhibitor of nitrification, but because it is a gas, it is difficult to add and keep in soil at the correct concentration to inhibit the oxidation of ammonium. Calcium carbide coated with layers of wax and shellac has been used to provide a slow-release source of acetylene to inhibit nitrification (Mosier 1994). This technique has increased the yield or recovery of N in irrigated wheat, maize, cotton, and flooded rice. Another product, a polyethylene matrix containing small particles of calcium carbide and various additives to provide controlled water penetration and acetylene release, has been developed as an alternative slow-release source of acetylene. This matrix inhibited nitrification for 90 days and considerably slowed the oxidation for 180 days (Freney et al. 2000).

A new nitrification inhibitor 3,4-dimethylpyrazole phosphate (DMPP, trade name ENTEC) was developed by the German company BASF AG (Linzmeier et al. 2001). Inhibition was achieved for 28 to 70 days with applications of 0.5 to 1.5 kg DMPP ha^{-1}, depending on the amount of N applied. Reliable data on the use of nitrification inhibitors in different crops and regions are not available. Surveys of U.S. farmers indicate that at present about 9 percent of the national maize area is treated with nitrification inhibitors, and this proportion has remained unchanged in recent years (Christensen 2002).

Deep placement of urea super granules for rice is re-emerging as a management alternative in certain parts of Asia with the potential to increase crop yields and reduce N losses (Mohanty et al. 1999). Deep-placement methods are currently being adopted in Bangladesh and Vietnam, more than 20 years after the improved efficiency of this technology was demonstrated. The practice was not adopted more rapidly because of the lack of a ready supply of super granules, the additional labor required, and the difficulty of placing the granules in the correct location. Small-scale fabrication of village-level urea briquette compactors, however, led to a dramatic increase in the use of super granules. The Bangladesh Department of Agricultural Extension reports that deep placement of super granules increased from 2 ha in 1995–1996 to 400,000 ha in 2000–2001 and sales increased to 91,840 tons.

Site-specific Nitrogen Management

Site-specific N management is a term used to refer to management of N tailored to a particular cropping system and season to optimize the congruence of supply and demand of N. Depending on when decisions are made and what information is used, site-specific N management strategies can be (1) prescriptive, (2) corrective, or (3) a combination of both. These strategies can be used to manage N in cropping systems that may range from labor-intensive, small-scale farming to highly mechanized management of large production fields (Dobermann and Cassman 2002). N can be applied homogeneously to a whole field or, in the most advanced case, rates may vary over short distances to account for spatial variability in soil N supply and crop N demand. Technologies are emerging that allow increased NUE by following any of these three strategies but with different potential and utilizing different tools in different environments.

Prescriptive Nitrogen Management

In prescriptive N management, amount and timing of N applications are prescribed before planting based on the expected crop response to fertilizer N. Information about N supply from indigenous sources, the expected crop N demand, the expected efficiency of fertilizer N, and the expected risk (climate) is needed. Increasing NUE can be achieved by fine-tuning prescription algorithms to local conditions and by better field characterization of any of the components used in such equations. Some scope exists for improved prescription algorithms, particularly in areas with excessive N use at crop stages with low N demand or where current N recommendations are of a very general nature. Examples include improved N recommendations for rice that account for major differences in variety types, cropping season, crop establishment method (Dobermann and Fairhurst 2000), significant increases in NUE of irrigated wheat in Mexico through fine-tuning of split applications (Riley et al. 2003), and improved N fertilizer management with vegetables in China (see Balasubramanian et al., Chapter 2, and Peoples et al., Chapter 4, this volume). Figure 3.1 shows the type of changes that can be expected in response to fertilizer N when improvements are made to crops or crop management.

Precision farming technologies have been developed to vary N prescriptions spatially within a field, based on various sources of spatial information (e.g., maps of soil properties, terrain attributes, on-the-go sensed electrical conductivity, remote sensing, yield maps). Fertilizer applications may be varied continuously, or a field is divided into few, larger subunits, commonly called *management zones*. In most studies, prescriptive variable-rate N fertilizer application reduced the average N rate required to achieve yields similar to those obtained with standard uniform management (Table 3.1). The specific technologies involved in all these steps are available, but they are not yet widely used, mainly because of uncertainties about the accuracy and profitability of this approach. Invasive or noninvasive soil sensors for assessing soil organic matter and soil nitrate status are being developed (Adamchuk et al. 2004); however, the potential for using soil sensors in N management remains unclear. Most available sensors provide only indirect or shallow depth measurements, for which conversion algorithms must be developed to derive N prescriptions. Another, perhaps more promising direction is the use of soil-crop simulation models for making N prescriptions at the field scale (Booltink et al. 2001; Table 3.1).

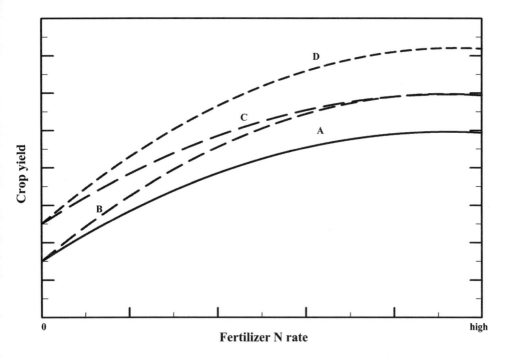

Figure 3.1. Generalized changes in crop yield response to fertilizer nitrogen (N) application as affected by improvements in crops or crop management. (A) Average N response function with low to medium fertilizer N efficiency. (B) Shift in the curvature (slope) of the N response function resulting from increased fertilizer N efficiency resulting from improved management. Measures to achieve this can include improved general crop management (e.g., plant density, irrigation, pest control) or improved N management technologies (e.g., placement, timing, modified fertilizers, balanced fertilization). (C) Upward-shifted N response function i.e., increase in the intercept (yield at zero N rate) but no change in the curvature because there is no increase in fertilizer N efficiency. An increase in the 0-N yield may be due to an improved variety with greater N acquisition or greater internal N utilization, amelioration of constraints that restricted uptake of indigenous N, or any measures that increase the indigenous N supply (crop rotation, application of organic materials). (D) Shift in the intercept and curvature of the N response function (i.e., increase in both 0-N yield and slope through a combination of measures). Full exploitation of yield potential is achieved by implementation of a site-specific, integrated crop management approach in which an advanced genotype is grown with near-perfect management, closely matching crop N demand and supply. As a result, both profit and fertilizer N use efficiency are high.

Table 3.1. Examples of different forms of prescriptive (p) or corrective (c) site-specific nitrogen management strategies implemented in field or on-farm studies

Crop, location	N treatment[1]	Decision tools[2]				N applied kg ha^{-1}	Yield t ha^{-1}	NUE[3] kg kg^{-1}
		S	M_r	M_t	D			
Maize, NE, USA[4]	Conventional	x	-	-	-	142	10.3	73
	Site-specific 1 (p)	x	-	-	-	141	10.4	74
	Site-specific 2 (p)	x	-	-	-	113	10.2	90
Maize, CO, USA[5]	Conventional	x	-	-	-	152	12.8	84
	Site-specific 1 (p)	x	-	-	-	163	12.4	76
	Site-specific 2 (p)	x	-	-	-	109	12.9	118
Wheat/triticale, Germany[6]	Conventional	x	-	-	-	175	9.2	53
	Site-specific (p)	x	-	-	-	166	9.1	55

(continued)

Modified from Dobermann et al. 2004.

[1] Conventional: Uniform N rate and fixed splitting of N (existing best management recommendations or farmers' practice); site-specific: Various approaches of more knowledge-intensive N management at different scales and using different decision tools.

[2] Decision tools used in N management: S—assessment of soil N supply using soil sampling or other techniques; M_r—soil/crop model to predict N rate; M_t—soil/crop model to predict splitting/timing of N applications; D—in-season diagnosis and adjustment of plant N using sensing tools (hand-held, on-the-go, or remote sensing).

[3] N use efficiency expressed as partial factor productivity = kg grain per kg N applied.

[4] Irrigated, average of 13 site years. Site-specific 1: Variable N rates based on a standard N recommendation algorithm using a uniform yield goal and grid maps of soil nitrate and soil organic matter. Site-specific 2: Reduced variable N rate, 15 to 25 percent less than site-specific 1.

[5] Irrigated, one site, 2 years. Site-specific 1: Variable N rates based on a standard N recommendation algorithm using a uniform yield goal and grid maps of soil nitrate and soil organic matter. Site-specific 2: Variable N rates based on a standard N recommendation algorithm using a variable yield goal and soil nitrate and soil organic matter data sampled by management zones.

[6] Two sites in 2002. Both N approaches included three N applications. Site-specific: Variable N rates adjusted according to management zones with different expected yield and soil characteristics.

CORRECTIVE NITROGEN MANAGEMENT

Because optimum N rates vary spatially and with seasonal conditions, corrective N management methods employ diagnostic tools to assess soil or crop N status during the growing season as the basis for making decisions about N applications at certain growth stages (Schroeder et al. 2000). Several promising technologies have emerged in recent years, with particular emphasis given to real-time measurements of crop greenness using tools such as near-infrared leaf N analysis, chlorophyll meters, leaf color charts, hand-held or on-the-go crop canopy reflectance sensors, or remote sensing (see Table

Table 3.1. *(continued)*

Crop, location	N treatment[1]	S	M_r	M_t	D	N applied kg ha^{-1}	Yield t ha^{-1}	NUE[3] kg kg^{-1}
		Decision tools[2]						

Crop, location	N treatment[1]	Decision tools[2]				N applied kg ha^{-1}	Yield t ha^{-1}	NUE[3] kg kg^{-1}
		S	M_r	M_t	D			
Wheat, UK[7]	Conventional	x	-	-	-	174	7.4	43
	Site-specific (p)	x	-	-	-	155	7.2	46
Wheat, OK, USA[8]	Conventional	-	-	-	-	90	2.1	23
	Site-specific (c)	-	-	-	x	109	2.3	21
Wheat, Germany[9]	Site-specific 1 (c)	x	-	-	x	178	6.3	35
	Site-specific 2 (p)	x	x	x	-	138	6.3	46
Rice, India[10]	Conventional	-	-	-	-	120	5.5	46
	Site-specific (c)	-	-	-	x	90	5.6	62
Rice, China[11]	Conventional	-	-	-	-	171	6.0	37
	Site-specific (p, c)	x	x	-	x	126	6.4	52

[7] Average of six site years. Site-specific: Variable N adjusted to management zones with different expected yield based on the mapped yield history.

[8] Dryland, average of four sites, 2001. Conventional: 45 kg N ha^{-1} preplant + 45 kg N ha^{-1} midseason. Site-specific: 45 kg N ha^{-1} preplant + variable sensor-based midseason amount at 1-m spatial resolution.

[9] One site, 2 years. Site-specific 1: Soil test-based preplant N + two variable rate applications using on-the-go Hydro N sensor. Site-specific 1: HERMES simulation model used for determining grid-cell specific N recommendations.

[10] One site, average of two varieties and 2 years. Site-specific: No pre-plant N, field-specific post-emergence N doses based on weekly chlorophyll meter readings using a SPAD threshold of 37.5 (Singh et al. 2002).

[11] Irrigated, average of 21 sites × 6 consecutive rice crops grown in Zhejiang Province, China. Conventional: Farmers' fertilizer practice. Site-specific: Field-specific NPK rates predetermined using a simple soil–crop model; in-season adjustment of N rates at key growth stages using a chlorophyll meter.

3.1). Significant increases in NUE are often achieved through reductions in N use by about 10 to 30 percent, whereas increases in yield tend to be small. These studies have also demonstrated that simple tools such as the leaf color chart developed for rice can result in improvements in NUE in smallholder farms of similar magnitude to those obtained with high-tech, large-scale approaches. A key issue for more widespread adoption is that of uncertain profitability of corrective N management approaches, particularly when the full costs of technology and risk are taken into account. Moreover, crop greenness is affected by numerous factors other than N, and sensing can be done only after the crop has developed enough biomass. Both N excess and deficiency may occur during early vegetative growth, which cannot be corrected with late-season N applications. Efforts are also ongoing to develop sensors and corrective strategies for managing crop quality, for example, grain protein yield in wheat (*Triticum aestivum* L.).

COMBINED APPROACHES

Integrating prescriptive and corrective concepts for quantifying how much, where, and when N must be added offers benefits. Uncertainties are reduced because a variety of information sources is used, including preseason assessment of soil N supply and in-season assessment of crop N needs. This strategy has been successfully used in field-specific nutrient management in irrigated rice, resulting in significant increases in yield, NUE, and profit across a large number of farms in Asia (Dobermann et al. 2002). Key components of this approach were measurement of grain yield in nutrient omission plots to obtain field-specific estimates of the indigenous supply of N, P, and K, a simple model for prescribing both nutrient requirements and the optimal amount of N to be applied before planting, and in-season upward or downward adjustments of predetermined N topdressings at critical growth stages based on actual chlorophyll meter or leaf color chart readings. Approaches are also emerging in which soil-crop simulation models are used in combination with field information and actual weather data (van Alphen and Stoorvogel 2000) to make N prescriptions at the beginning of the growing season as well as in real-time during crop growth. In-season prediction of crop yield potential using models is becoming available for cereals (Bannayan et al. 2003) and offers new possibilities for real-time N management in prescriptive-corrective concepts.

Conservation Agriculture

Conservation agriculture is basically a management system that embodies zero or minimum tillage with direct seeding, retention of crop residues, and the maintenance of soil cover with crops and crop rotations.

CONSERVATION TILLAGE IN U.S. MAIZE–SOYBEAN SYSTEMS

Conservation tillage is used on nearly 40 percent of the land in maize (*Zea mays* L.) production in the United States. Requirements for N in no-till systems differ from those in tilled systems and, depending on how N is managed, NUE may be either lower or higher than with tillage. At present, no significant differences have been found among tillage systems in terms of N rates used by U.S. farmers, timing of N application, or tools for N management (Christensen 2002). With increasing adoption of conservation tillage practices, the need for developing N management strategies and technologies that are fine-tuned to the specific requirements of these systems is increasing. It is also likely that breeding specifically for no-tillage could increase N use efficiency by matching fine root distribution better to the altered distribution of soil organic matter (while ensuring that sufficient roots are available at such a depth that drought susceptibility is not increased).

CONSERVATION TILLAGE IN RICE–WHEAT SYSTEMS IN ASIA

Rice and wheat are grown in rotation on 17.5 million ha of land in Asia, providing food for about one billion people (Ladha et al. 2003). In the last decades, rapid growth in the annual production of wheat (3.0 percent) and rice (2.3 percent) has kept pace with population growth; however, problems such as excessive and unbalanced use of N fertilizer, poor crop residue management, and groundwater depletion are emerging.

Adoption of resource-conserving technologies can increase the NUE and that of water and energy in rice–wheat systems. The techniques used include permanent direct-seeded and bed-planted rice, surface seeding systems, laser-leveling of irrigated fields, management of crop residues for surface cover, and reducing rice fallows. No-till wheat after rice, now covering 0.5 m ha in the Indo-Gangetic Plain, led to a range of benefits to farm families, among them substantial improvements in farm-level water productivity (Ladha et al. 2003). Recently attempts were made to grow rice with reduced tillage under aerobic conditions. When N fertilizer was deep-placed using zero-tillage drills at the time of seeding, yields were maintained and NUE increased.

Communication and Dissemination of Emerging Technologies

The implementation of new ideas and approaches to enhance NUE depends on the flow of information about new developments to farmers and the availability of these technologies. Often farmers are key actors in the innovation, development, and testing of technologies; irrespective of who develops new knowledge, efficient communication methods are required for "upscaling" innovations. A general lesson, initially emphasized in tropical countries and gaining increasing ground in Europe and Australia, is the use of "participatory" learning approaches in which farmers play a strong role in innovation, development, and testing of technologies together with researchers.

Innovation in Communication Approaches

Public-sector extension organizations are no longer a favored model for education and extension of emerging technologies and new knowledge throughout the world. In developing countries, government extension services are generally moribund because of chronic underfunding and undercapacity. In more developed countries, the role of extension is increasingly seen as a service for which farmers should pay, leading to privatization of government extension and applied research organizations. In India and Vietnam, the private sector is developing networks of agri-service centers that represent an exciting and effective development in communication and extension. These privately funded centers offer educational programs (e.g., child care, human nutrition) for women farmers and have drawn the attention of the Sustainable Development Movement. Trained staff members provide complete production packages that include fertilizers, seeds, and crop-protection products (as well as training in their safe use and protective clothing), and farmers obtain advice on best management practices tailored to their specific conditions and assistance in obtaining credit. In addition, the Indian Council of Agricultural Research is developing a comprehensive Internet-based information system, with access points in the villages to allow direct access by the farmers to local recommendations. In Latin America, the Internet is being increasingly used to enable farmers to access information (see www.ciat.org). Other initiatives, such as the market information systems developed by the International Fertilizer Development Center, assist farmers in West Africa in decision-making relating to choice of crops and profitability of inputs and other information.

Computer-based Decision Support Systems

Because many combinations of farming systems and management practices exist, it is difficult to study all possible combinations. A number of computer-simulation models have been developed to assess the impacts of management practices on NUE and N loss, and a few examples are given later. The Nitrate Leaching and Economics Analysis Package (NLEAP) has been used to predict nitrate dynamics and NEU for cropping systems with different rooting depths (Delgado et al. 1998). A nutrient budget model, OVERSEER, was developed in New Zealand with the aims of providing reasonable estimates of inputs and outputs of N and examining management practices which reduce loss of N (Ledgard et al. 2001).

The APSIM crop-simulation model has been used together with farmers in Australia to assist in combined prescriptive and corrective N fertilizer management. Video links allow groups of farmers to discuss regularly the season's progress, with researchers running simulation-modeling scenarios interactively in the Farmers Advisors Researchers Simulation Communications and Performance Evaluation (FARMSCAPE) approach (McCown 2002).

Other Approaches to Decision Support

Decision trees to guide the use of combinations of organic nutrient resources and N fertilizers have been developed (Vanlauwe et al., Chapter 8, this volume). Simple field assessments based on leaf color (N content), toughness (a surrogate for lignin), and astringency on taste (reactive polyphenol agents) have allowed these decision trees to be translated into forms that can be used for discussion between farmers and development workers (Giller 2000).

Although most of this chapter focuses on N management at the field scale, resource-limited farmers make decisions about allocating the N fertilizers they can obtain at the farm scale. Strong gradients of soil fertility exist, with more fertile fields generally found close to the homesteads (Vanlauwe et al., Chapter 8, this volume). The agronomic NUE may vary from more than 40 kg grain kg N applied^{-1} on the more fertile fields to fewer than 4 kg grain kg N applied^{-1} on degraded outfields within the same smallholder farm in Zimbabwe (S. Zingore, personal communication). The very poor agronomic efficiencies on the degraded soils are due to multiple nutrient deficiencies and critically low soil organic-matter contents leading to problems of water availability. Therefore, significant inputs of organic matter combined with fertilizers are needed to bring such fields back into productive agriculture. Fertilizer use in such complex and spatially heterogeneous farming systems requires understanding of these interactions, and national or regional static fertilizer recommendations are clearly inappropriate. Further, access to purchased inputs depends on the resource endowment status of the farmers. Nutrient requirements for such systems need to be targeted to the crop rotation, as use of animal manures may allow growth of N_2-fixing legumes on such soils as a means to raising the soil fertility status for cereal production (Vanlauwe et al., Chapter 8, this volume). With investment of inputs for crop production between fields come signifiant tradeoffs, demonstrating that understanding returns to N fertilizer use necessitates farm-scale analysis.

Decision making relating to environmental targets for water quality requires consideration of N use and NUE at an even larger scale, such as the watershed or the total area that contributes to specific groundwater aquifers (Peoples et al., Chapter 4, this volume).

Conclusions

Important prerequisites for the adoption of advanced N management technologies are that they must be simple, provide consistent and large enough gains in NUE, involve little extra time, and be cost-effective. Many emerging technologies do not automatically fulfill all these criteria and may require some initial support for adoption. A key issue is that the risk of profit loss must be small and, in many cases, that profit increase must be substantial to make the technology attractive for a farmer. This can be achieved in two ways: First, if the new technology leads to a small increase in crop yield with the same amount or less N applied than the conventional practice, the resulting increase in profit is usually sufficiently attractive for a farmer, particularly in developing countries or large-scale grain farms in North and South America or in Australia, where there is still potential and a need to produce more food and feed. Second, where yield increases are difficult to achieve, where increasing crop yield is of less priority, or where reducing the creation of reactive N in agriculture is the top societal priority, adoption of new technologies that increase NUE but have little effect on farm profit needs to be supported by appropriate technology incentives. An example for this is agriculture in many countries of Western Europe.

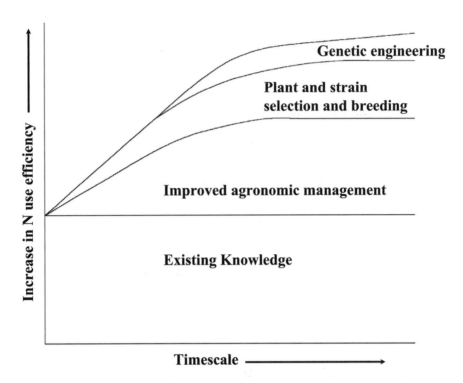

Figure 3.2. The likely impact of research investment in increasing nitrogen use efficiency.

In Figure 3.2, we have indicated where we expect the greatest gains in NUE to be realized in the future. Changing socioeconomic or agroecological conditions, or policies and measures that target economic or environmental goals, will have an overriding influence on whether new and emerging technologies for NUE will gain acceptance by farmers and society (Palm et al., Chapter 5, this volume).

Literature Cited

Adamchuk, V. I., J. W. Hummel, M. T. Morgan, and S. K. Upadhaya. 2004. On-the-go soil sensors for precision agriculture. *Computers and Electronics in Agriculture* (in press).

Bannayan, M., N. M. J. Crout, and G. Hoogenboom. 2003. Application of the CERES-Wheat model for within-season prediction of winter wheat yield in the United Kingdom. *Agronomy Journal* 95:114–125.

Bänziger, M., and M. Cooper. 2001. Breeding for low input conditions and consequences for participatory plant breeding: Examples from tropical maize and wheat. *Euphytica* 122:503–519.

Booltink, H. W. G., B. J. van Alphen, W. D. Batchelor, J. O. Paz, J. J. Stoorvogel, and R. Vargas. 2001. Tools for optimizing management of spatially variable fields. *Agricultural Systems* 70:445–476.

Cassman, K. G., A. Dobermann, D. T. Walters, and H. Yang. 2003. Meeting cereal demand while protecting natural resources and improving environmental quality. *Annual Review of Environmental Resources* 28:315–358.

Christensen, L. A. 2002. *Soil, nutrient, and water management systems used in U.S. corn production.* Agriculture Information Bulletin no. 774. Washington, D.C.: Economic Research Service, United States Department of Agriculture.

Delgado, J. A., M. J. Shaffer, and M. K. Brodahl. 1998. New NLEAP for shallow and deep rooted crop rotations. *Journal of Soil and Water Conservation* 53:338–340.

De Melo, P. E. 2003. The root systems of onion and *Allium fistulosum* in the context of organic farming: A breeding approach. PhD thesis, Wageningen University, The Netherlands.

Dobermann, A., and K. G. Cassman. 2002. Plant nutrient management for enhanced productivity in intensive grain production systems of the United States and Asia. *Plant and Soil* 247:153–175.

Dobermann, A., and T. H. Fairhurst. 2000. *Rice: Nutrient disorders and nutrient management.* Singapore: Potash and Phosphate Institute, Makati City: International Rice Research Institute.

Dobermann, A., S. Blackmore, S. E. Cook, and V. I. Adamchuk. 2004. Precision farming: Challenges and future directions, in *New directions for a diverse planet.* Proceedings of the 4th International Crop Science Congress, 26 September–1 October 2004, Brisbane, Australia. Published on CD ROM. Web site www.regional.org.au/aw/cs.

Dobermann, A., C. Witt, D. Dawe, G. C. Gines, R. Nagarajan, S. Satawathananont, T. T. Son, P. S. Tan, G. H. Wang, N. V. Chien, V. T. K. Thoa, C. V. Phung, P. Stalin, P. Muthukrishnan, V. Ravi, M. Babu, S. Chatuporn, M. Kongchum, Q. Sun, R. Fu, G. C. Simbahan, and M. A. A. Adviento. 2002. Site-specific nutrient management for intensive rice cropping systems in Asia. *Field Crops Research* 74:37–66.

Drost, R., R. Koenig, and T. Tindall. 2002. Nitrogen use efficiency and onion yield increased with a polymer-coated nitrogen source. *Hortscience* 37:338–342.

Fashola, O. O., K. Hayashi, and T. Wakatsuki. 2002. Effect of water management and polyolefin-coated urea on growth and nitrogen uptake of indica rice. *Journal of Plant Nutrition* 25:2173–2190.

Freney, J. R., P. J. Randall, J. W. B. Smith, J. Hodgkin, K. J. Harrington, and T. C. Morton. 2000. Slow release sources of acetylene to inhibit nitrification in soil. *Nutrient Cycling in Agroecosystems* 56:241–251.

Giller, K. E. 2000. Translating science into action for agricultural development in the tropics: An example from decomposition studies. *Applied Soil Ecology* 14:1–3.

Giller, K. E., and G. Cadisch. 1995. Future benefits from biological nitrogen fixation: An ecological approach to agriculture. *Plant and Soil* 174:255–277.

Giller, K. E., and R. Merckx. 2003. Exploring the boundaries of N_2-fixation in non-legumes: An hypothetical and experimental framework. *Symbiosis* 35:3–17.

Greenwood, D. J., G. Lemaire, G. Gosse, P. Cruz, A. Draycott, and J. T. Neeteson. 1990.

Decline in percentage N of C3 and C4 crops with increasing plant mass. *Annals of Botany* 66:425–436.

Hubbell, D. H. 1995. Extension of symbiotic biological nitrogen fixation technology in developing countries. *Fertilizer Research* 42, 231–239.

Kundu, D. K., and J. K. Ladha. 1997. Effect of growing rice on nitrogen mineralization in flooded soil. *Soil Science Society of America Journal* 61:839–845.

Ladha J. K., and P. M. Reddy. 2000. *The quest for nitrogen fixation in rice.* The Philippines: International Rice Research Institute.

Ladha, J. K., J. E. Hill, J. D. Duxbury, R. K. Gupta, R. J. Buresh. 2003. *Improving the productivity and sustainability of rice-wheat systems: Issues and impact.* American Society of Agronomy Spec. Publ. 65. Madison, Wisconsin: ASA, CSSA, SSSA.

Ledgard, S. F., B. S. Thorrold, R. A. Petch, and J. Young. 2001. Use of OVERSEER as a tool to identify management strategies for reducing nitrate leaching from farms around Lake Taupo, in *Precision tools for improving land management.* Edited by L. D. Currie and P. Loganathan. Occasional report no. 14. Palmerston North, New Zealand: FLRC, Massey University.

Linzmeier, W., R. Gutser, and U. Schmidhalter. 2001. Nitrous oxide emission from soil and from a nitrogen-15- labelled fertilizer with the new nitrification inhibitor 3,4, dimethylpyrazole phosphate (DMPP). *Biology and Fertility of Soils* 34:103–108.

McCown, R. L. 2002. Changing systems for supporting farmers' decisions: Problems, paradigms, and prospects. *Agricultural Systems* 74:179–220.

Mohanty, S. K., U. Singh, V. Balasubramanian, and K. P. Jha. 1999. Nitrogen deep-placement technologies for productivity, profitability, and environmental quality of rainfed lowland rice systems. *Nutrient Cycling in Agroecosystems* 53:43–57.

Mosier, A. R. 1994. Nitrous oxide emission from agricultural soils. *Fertilizer Research* 37:191–200.

Peng, S., and K. G. Cassman. 1998. Upper thresholds of nitrogen uptake rates and associated nitrogen fertilizer efficiencies in irrigated rice. *Agronomy Journal* 90:178–185.

Riley, W. J., I. Ortiz-Monasterio, and P. A. Matson. 2003. Nitrogen leaching and soil nitrate, nitrite, and ammonium levels under irrigated wheat in Northern Mexico. *Nutrient Cycling in Agroecosystems* 61:223–236.

Schroeder, J. J., J. J. Neeteson, O. Oenema, and P. C. Struik. 2000. Does the crop or the soil indicate how to save nitrogen in maize production? Reviewing the state of the art. *Field Crops Research* 66:151–164.

Shaviv, A. 2000. Advances in controlled release fertilizers. *Advances in Agronomy* 71:1–49.

Shoji, S., and H. Kanno. 1995. Innovation of new agrotechnology using controlled release fertilizers for minimizing environmental deterioration, in *Proceedings of the Dahlia Gredinger Memorial International Workshop on Controlled/Slow Release Fertilizers*, edited by Y. Hagin, et al. Haifa, Israel: Technion.

Shoji, S., J. Delgado, A. Mosier, and Y. Miura. 2001. Use of controlled release fertilizers and nitrification inhibitors to increase nitrogen use efficiency and to conserve air and water quality. *Communications in Soil Science and Plant Analysis* 32:1051–1070.

Singh, B., Y. Singh, J. K. Ladha, K. F. Bronson, V. Balasubramanian, J. Singh, and C. S. Khind. 2002. Chlorophyll meter- and leaf color chart-based nitrogen management for rice and wheat in northwestern India. *Agronomy Journal* 94:821–829.

Tirol-Padre, A., J. K. Ladha, U. Singh, E. Laureles, G. Punzalan, and S. Akita. 1996.

Grain yield performance of rice genotypes at sub-optimal levels of soil N as affected by N uptake and utilization efficiency. *Field Crops Research* 46:127

Trenkel, M. E. 1997. *Controlled release and stabilized fertilizers in agriculture*. Paris: IFA.

van Alphen, B. J., and J. J. Stoorvogel. 2000. A functional approach to soil characterization in support of precision agriculture. *Soil Science Society of America Journal* 64: 1706–1713.

4

Pathways of Nitrogen Loss and Their Impacts on Human Health and the Environment

Mark B. Peoples, Elizabeth W. Boyer, Keith W. T. Goulding, Patrick Heffer, Victor A. Ochwoh, Bernard Vanlauwe, Stanley Wood, Kazuyuki Yagi, and Oswald van Cleemput

The contribution of fertilizer nitrogen (N) to total N inputs into agricultural systems rose from just 7 percent in 1950 to 43 percent by 1996 (Mosier 2001). This increase impacted food production in two main ways (Crews and Peoples 2004). First, the availability of synthetic fertilizers provided a relatively cheap and convenient means for farmers to meet plant demands for N throughout the growing season, and about 40 percent of the observed increases in grain yield since the 1960s have been attributed directly to N fertilizer (Brown 1999). Second, the use of fertilizers allowed farmers to grow cereals or other crops on land that would otherwise have been dedicated to the fertility-generating phase of a rotation sequence. Before the advent of N fertilizers, 25 to 50 percent of a farm was typically maintained in a legume-rich pasture or cover crop (Smil 2001).

Society has gained considerable benefits from the additional food production achieved with the widespread adoption of N fertilizers (Wood et al., Chapter 18, this volume). For example, an estimated 40 percent of the protein consumed globally by humans originated from N supplied as fertilizer (Smil 2001). Unfortunately, often less than 50 to 60 percent of the N applied to crops or pastures might be recovered by plants under current farming practices (Balasubramanian et al., Chapter 2, this volume). Some of the inefficiencies in uptake can be attributed to the volatile and mobile nature of N. It is easily transformed among various reduced and oxidized forms and is readily distributed by hydrologic and atmospheric transport processes. Nitrogen can be lost from the site of application in farmers' fields through soil erosion, runoff, or leaching of nitrate or dissolved forms of organic N or through gaseous emissions to the atmosphere in the forms of ammonia (NH_3), nitrogen oxides (NO and NO_2), nitrous oxide (N_2O), or dinitrogen (N_2) (Goulding, Chapter 15, this volume). All these avenues of loss, with the important exception of N_2, can potentially impact on one or more environmental hazards or have important implications for human health.

Fertilizer Nitrogen in Context

Many of the environmental effects described in the following sections are functions of the total net N inputs to a region. Nitrogen sources that are not intentional but that occur as a result of human activities or natural processes include atmospheric N deposition, human waste, and natural biological N_2 fixation in noncultivated vegetation such as forests. Other N inputs are deliberate and managed in agricultural lands. In addition to mineral fertilizer N, other sources of N include biological N_2 fixation by legumes and other symbiotic or associative relationships between microorganisms and plants (Peoples 2002) and the applications of manures, compost, crop residues or other organic materials (Boyer et al., Chapter 16, this volume).

The relative importance of these different N sources varies greatly by region and is related to a range of socioeconomic factors that include population density and patterns of land use. When considering total intentional N inputs to agricultural lands, fertilizer N inputs are higher than other managed N inputs in Asia (by 100 percent), Europe (by 70 percent), and North America (by 40 percent). In contrast, managed inputs to agricultural lands from manure and biological N_2 fixation dominate over synthetic fertilizer inputs in Africa (by 40 percent), Latin America (by 80 percent), and Oceania (by 60 percent) (Boyer et al., Chapter 16, this volume).

One important question to consider is whether losses of N from synthetic, mineral fertilizer sources differ from those of organic origin such as manure. Unfortunately only a limited number of comparisons of different systems have been done using ^{15}N-labeled inputs that allow direct measurement of plant uptake and soil retention of the applied N and provide indirect information about losses (generally based on the amount of the applied ^{15}N not recovered in either the plant or soil, Table 4.1). When inputs are properly managed, crops in rain-fed systems usually recover more applied N from fertilizer than from organic inputs, but a higher proportion of the applied N generally remains in the soil at harvest with organic sources. The range of estimated losses from both sources, therefore, is often rather similar (Table 4.1). The situation seems to be somewhat different in lowland rice or irrigated systems, where losses from fertilizer N can be substantially higher than losses of applied organic N (Table 4.1). These observations should be qualified, however, by acknowledging that (1) it is not clear how many of the comparative studies summarized in Table 4.1 have used "best management practices" when applying the fertilizer; and (2) often the ^{15}N labeled legume inputs represented only shoot material, which ignores the potentially large contributions of belowground N in legume-based rotations associated with legume roots and nodules (Rochester et al. 2001).

Table 4.1. Examples of the fate of nitrogen in field experiments involving the application of ^{15}N-enriched fertilizers or legume residues, indicating the range estimates of the recovery and losses of applied N

Source of N applied	Crop uptake (% applied N)	Recovery in soil (% applied N)	Total recovered [crop + soil] (% applied N)	Unrecovered [assumed lost] (% applied N)
Rain-fed cereal cropping [1]				
Fertilizer	16–51	19–38	54–84	16–46
Legume	9–19	58–83	64–85	15–36
Irrigated cotton [2]				
Fertilizer	—	—	4–17	83–96
Legume	—	—	62–82	18–38
Lowland rice [3]				
Fertilizer	—	—	61–65	35–39
Legume	—	—	87–93	7–13

[1] Wheat data from Canada (Janzen et al. 1990) and Australia (Ladd and Amato 1986); maize and barley data from the United States (Harris et al. 1994), and maize data from Africa (Vanlauwe et al. 1998, Vanlauwe et al. 2001a).

[2] Data derived from Rochester et al. (2001).

[3] Data derived from Diekmann et al. (1993) and Becker et al. (1994).

Table 4.2. Estimates of annual global gaseous emissions of N_2O, NO, and NH_3 from nitrogen fertilizer or manures applied to crops and grasslands [1]

Amount N applied (million t N)	N_2O (million t N)	NO (million t N)	NH_3 (million t N)
As fertilizer N			
77.8	0.9	0.6	11.2
	(1.2% of N applied)	(0.8% of N applied)	(14.4% of N applied)
As manure			
32.0	2.5	1.4	7.8
	(7.8% of N applied)	(4.4% of N applied)	(24.4% of N applied)

[1] Data collated for 1436 million ha of crops and 625 million ha of grasslands receiving applications of either fertilizer N or manure in 1995 (IFA/FAO 2001). Values were derived from extrapolations of research results, using statistical data, geographic information, and assumptions about fertilizer management. It is important to note that the flux estimates provided here account only for the increased direct emissions resulting from the addition of synthetic fertilizer or livestock manure. The estimates do not account for further emissions that might subsequently result from nitrate leaching, runoff, and ammonia volatilization.

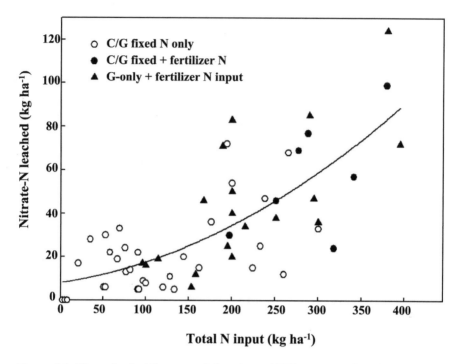

Figure 4.1. Nitrate leached from grazed clover/grass (C/G) or grass-only (G) pastures as affected by annual aboveground inputs of N from legume N_2 fixation or applications of fertilizer N. Data derived from studies undertaken in Australia, New Zealand, France, and the United Kingdom collated by Fillery (2001) and Ledgard (2001).

Derived estimates of global emissions of gaseous N products (Table 4.2) do imply key differences in N losses between fertilizer N and manure sources of N applied to crops and grasslands. Although the absolute amounts of N calculated to be lost directly as N_2O, NO, or NH_3 from fertilizer and manures do not appear to differ greatly, differences are clear in the extent of N losses when expressed as a proportion of N applied (Table 4.2). Direct measurements of the amounts of N leached from grazed temperate pastures on the other hand suggest that the amounts lost are more a function of the size of the annual input of N than whether the source of N was derived from biological N_2 fixation or fertilizer (Figure 4.1).

Factors Controlling Nitrogen Loss Processes

Our subsequent discussion of loss processes, the factors controlling them, and their impacts are based on Figure 4.2, which shows the interactions between N input and N loss processes. We refer to the input of N to a field, plant uptake and off-take of N in

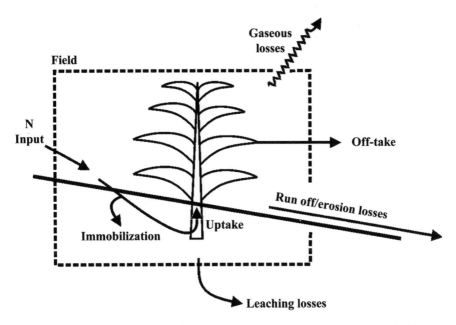

Figure 4.2. Schematic diagram indicating the interactions between N input and N loss processes.

agricultural produce and residues, local immobilization of N by soil microbes, and gaseous, leaching, or runoff/erosion losses.

Table 4.3 provides a simplified summary of the key factors involved and indicates the complexity of N loss processes. The controlling factors are divided into environmental variables, which are largely uncontrollable, and the impacts of human activity through which we have some ability to manage losses. Volatilization, leaching, runoff, and erosion are local losses, but they may have an offsite impact nearby.

Denitrifrication is clearly the most complex process and the one most influenced by environmental variables (Table 4.3). One aspect of its complexity is the proportion of emissions as N_2O, which has important environmental consequences, or as N_2, which has no adverse implications (Peoples et al. 1995). The influence of different variables on the ratio of N_2O: N_2 in gaseous emissions is illustrated in Table 4.4. Because human activity can influence almost every process listed in Table 4.4, control is possible but is likely to be complicated. The most important factors would appear to be N inputs, stocking rates of grazing animals, and land use change (Table 4.3).

There is little doubt that the relative importance of the various loss processes will vary considerably across different regions based on climate, soils, dominant land use, and sources of N inputs used for agriculture (Goulding, Chapter 15, this volume). Estimates of the contributions of different countries or regions to the total global losses

Table 4.3. Summary of key processes and factors influencing nitrogen loss[1]

Factors	Nitrification	Denitrification	Volatilization	Leaching	Runoff and erosion
Environmental variables					
Microbial activity	xxx	xxx	—	—	—
Soil pH	xx	x	xxx	—	—
Salinity	—	—	xxx	—	—
Topsoil texture	—	xxx	xx	xxx	xxx
Soil profile	—	xxx	xxx	xxx	xx
Soil aeration	xxx	xxx	—	—	—
Temperature	xx	xx	xx	—	—
Water supply	xx	xxx	xx	xxx	xxx
Available C	xx	xxx	—	—	x
Topography	—	—	—	—	xxx
Impact of human activity					
N inputs, type	xx	xx	xxx	xxx	—
Amount	xxx	xxx	xxx	xxx	xxx
Placement	x	xx	xxx	x	xx
Timing	xx	xx	xx	xx	xx
Plant spp/variety	x	x	—	xxx	xx
Residues, quality	xxx	xxx	x	x	xx
Groundcover	xx	xxx	x	xx	xxx
Tillage	x	xx	x	x	xxx
Soil compaction	x	xxx	—	x	xxx
Drainage	xx	xxx	x	xxx	xx
Irrigation	xx	xxx	xx	xxx	xxx
Stocking rate	xx	xx	xxx	xxx	xx
Land use change	xxx	xxx	xxx	xxx	xxx

[1] Each factor is ranked against each process according to its relative importance in controlling that process. The symbol "x" represents relative standard: small (x), medium (xx), high (xxx), little or no (—) importance. These rankings include both positive and negative effects. Each factor is considered entirely separately from the others. For example, in the field, water supply influences soil aeration, but such interactions are ignored here. Topsoil texture is separated from soil profile because the former has a specific effect on biological processes, whereas the latter influences physical properties, such as hydrology. Available C is not just an environmental variable, but it is also influenced by human activity. Although topography is a multidetermining factor, the rankings above refer only to slope. Nitrogen inputs include mineral fertilizer, manures, compost, and biological N_2 fixation but not atmospheric deposition. Within tillage, *no till* does not include groundcover, and *stocking rate* is used as a more general term than *grazing intensity.*

Table 4.4. Factors influencing the ratio of $N_2O:N_2$ emissions [1]

Factor	Variable	Ratio
Single parameters		
Nitrate, nitrite	Increasing concentration	Increase
Anoxicity	Increasing O_2	Increase
Temperature	Decreasing	Increase
Sulphide	Increasing	Increase
Available C	Increased availability	Decrease
pH	Increasing	Decrease
Combined parameters		
Plants	Increased presence	Decrease
Soil depth	Deeper	Decrease
Drying/wetting	Prolonged period	Decrease
Moisture	Increased	Decrease
Denitrification rate	Increased	Decrease

[1] van Cleemput (1998).

of N in a gaseous (e.g., IFA/FAO 2001) or liquid phase (e.g. Van Drecht et al. 2003) are based on a wide range of assumptions and extrapolations from research findings and point source measurements. Clearly, such derived estimates are likely to be more reliable for those regions and countries that are most "data rich"; given the technical difficulties in measuring the different pathways of N losses, it is inevitable that more quantitative information at different levels of resolution will be available for some loss processes than for others and in some regions more than others (Goulding, Chapter 15, this volume). Yet even comprehensive, coordinated investigations across countries and ecosystems still need to address a number of potential methodologic difficulties in upscaling research data collected from enclosures and small plots to the field, farm, landscape, or regional scale. The problems of both temporal and spatial scaling make the comparison of loss pathways across different scales extremely difficult (Goulding, Chapter 15, this volume).

Interactions with Other Factors

Soil biological processes depend heavily on soil organic C. Because soil N transformation processes are largely driven by biological activity, and organic C represents an energy source and source of nutrient supply, a strong interaction between the dynamics of C and N is expected. The availability of organic C is influenced by both the quantity of C and the quality of the organic material (e.g., C:N ratio, polyphenol, cellulose, or lignin contents; Kumar and Goh 2000). Both factors change within the soil profile, giving rise to different localized rates of N transformations. Applications of organic material to the soil or crop residues with high C:N ratios can stimulate microbial immobilization of NH_4 and NO_3 and hence restrict nitrification resulting from competition for substrate. Increased immobilization might reduce N_2O emissions or reduce the total losses of fertilizer N (Vanlauwe et al. 2002). In contrast, during decomposition of organic materials with low C:N ratios, a rapid release of NH_4 and NO_3 will occur, creating conditions suitable for the production of N_2O (Table 4.4). The addition of organic material to soil will greatly enhance microbial oxygen consumption so that oxygen deficiency may occur within localized zones and denitrification can occur even under aerobic conditions (Gök and Ottow 1986).

Interactions between N fertilizer and other growth factors, such as water, phosphorus, potassium, sulphur, or micronutrients, may alter the relative importance and magnitude of the various N loss processes. This can happen either directly by impacting on the major factors driving loss processes (see Table 4.3) or indirectly by improving crop growth and N uptake. By alleviating one of the constraints to crop growth other than N through application of, for example, phosphatic fertilizer, crops will exhibit a higher demand for N and consequently compete more strongly with the N loss processes. Besides application of specific nutrients in the form of fertilizer, organic resources can potentially influence soil-available phosphorus (Nziguheba et al. 2000), soil moisture conditions, cation status, or pest and disease dynamics (Vanlauwe et al. 2001b). The retention of plant residues or applications of other forms of organic matter to the soil surface can also substantially reduce erosion and runoff losses.

Potential Applications of Simulation Models

The process-based knowledge of N and C cycling has in numerous instances been integrated in mechanistic and dynamic simulation models. Such models offer the potential to analyze the contribution of individual components of a system to N cycling and losses. This is typically undertaken through sensitivity analyses and intermodel comparisons, which may be used to identify gaps in current process understanding. Modeling can also serve as a tool for interpreting experimental results and extrapolating to new environmental and management conditions (Smith et al. 1997). The available models often have different strengths in scale or loss pathways. Most models function at the plot or field scale (e.g., Hansen et al. 1991), whereas few models integrate interactions also at the farm scale (Berntsen et al. 2003). Many simulate nitrate leaching, some simulate denitrification and N_2O emissions (e.g., Parton et al. 2001), but only a few models simulate ammonia volatilisation (Sommer et al. 2003). Models are often applied for estimating losses at higher spatial or temporal scales; however, this often involves simplifying model inputs or model structure (e.g., Børgesen et al. 2001) and requires validation (Goulding, Chapter 15, this volume).

Environmental and Health Impacts

New inputs of reactive N entering regional landscapes have a "cascading" effect in a wide range of changes that impact on humans and ecosystems in different ways in various parts of the world (Galloway and Cowling 2002). Some of these changes are beneficial for society, particularly with regard to enhanced food production, although other consequences of nutrient enrichment are detrimental to terrestrial and aquatic ecosystems and to human health. Many of these effects do not occur in isolation and are linked through various biogeochemical processes. For example, adapting the "cascade model" of Galloway and Cowling (2002), a given reactive N atom passing from the atmosphere through a terrestrial, agricultural landscape might do the following:

• Increase ozone concentrations in the troposphere, decrease atmospheric visibility, and increase precipitation acidity.
• Be fixed and applied in the form of an inorganic fertilizer to enhance the productivity of agricultural systems.
• Be taken up in the harvested part of a crop or leached, increasing soil and water acidity.
• Be consumed by humans in food or water or digested and excreted by the human body, ending up in septic and sewer systems.
• Be consumed by animals in feed or water, digested and excreted as manure, or spread on the landscape.
• Be transported to fresh or coastal waters contributing to eutrophication.

In agricultural systems, most N inputs are deliberate and managed and might come from fertilizer, manure, or the cultivation of leguminous crops. The potential consequences of such N applications are summarized in the following sections under local (field) and off-site (product off-take, losses in gaseous or liquid forms) impacts as depicted in Figure 4.2. Only the most important effects are described in any detail. Further information on the implications of N use in agriculture can be found in numerous reviews covering environmental (e.g., Galloway et al. 1995) and human health (e.g., Townsend et al. 2003) issues.

Most of the direct and indirect effects of N applications identified in the following sections have obvious beneficial consequences associated with food production or deleterious consequences associated with atmospheric, terrestrial, and aquatic ecosystems. Human health threats posed by elevated nitrate levels in drinking water and foodstuff, however, are not well understood scientifically and thus remain controversial.

It is also important to realize that, in addition to the issues listed here, there may be other less apparent environmental implications associated with the use of fertilizer compared with alternative sources of N. Between 0.7 and 1.0 tonne of CO_2-C is released with every tonne of ammonia manufactured. About half the CO_2-C will be reused if that ammonia is subsequently converted to urea, but this CO_2 is still rapidly released to the atmosphere when the urea is applied in the field (Jenkinson 2001). Therefore, the additional global warming potential generated by the use of fossil energy to produce N fertilizers ideally should also be considered when undertaking a full inventory of environmental consequences (Crews and Peoples 2004).

Summary of Potential Field (Local) Effects

The impacts of fertilizer N in the field in which it is applied can be summarized as follows:

1. Assimilation (immobilization) of inorganic N by the soil microbial population
2. Changes in soil N storage and C sequestration
3. Soil acidification
4. Changes in land-use patterns
5. Potential health and safety risks associated with ammonium nitrate (explosive) and anhydrous ammonia.

Data collected from long-term trials in Brazil demonstrate that increased N inputs can provide environmental benefits by increasing soil C and N stocks and thus enhancing C sequestration (Sisti et al. 2004). On the other hand, increased N inputs can also contribute to one of the most insidious forms of soil degradation, namely, soil acidification. The addition of reduced, inorganic N to soils in certain fertilizers (urea or anhydrous ammonia) or following ammonification of organic matter (such as legume residues) does not directly lead to soil acidification. For these inputs to contribute to soil acidification, ammonium must undergo nitrification to form nitrate, and then the nitrate and associated cations must subsequently be leached down the soil profile (Kennedy 1992). In contrast, the application of ammonium-based fertilizers (ammonium nitrate, ammonium phosphate, or ammonium sulfate) increases the net H^+ concentration of soils and thus directly contributes to soil acidification, even in the absence of nitrate leaching (Kennedy 1992). Fertilizer applications can also have indirect effects on soil acidification because the resultant increased crop productivity leads to an enhanced rate of cation removal in agricultural produce. So although the acidification of soils is a natural process, it tends to be accelerated with increased N inputs. If allowed to proceed long enough (i.e., if the soil pH is not regularly corrected through the application of lime and the soil becomes acid), crop performance will ultimately be reduced as a result of aluminum and manganese toxicities and reduced availabilities of a range of nutrients (Crews and Peoples 2004).

Summary of Potential Product Off-take Effects

The following are the main effects of the off-take (removal in harvested produce) of N in crops:

1. Increased yields and nutritional quality of foods to satisfy dietary consumption and food preferences for growing human populations or of feedstuffs to meet animal nutrition requirements.
2. Possible threats or benefits to food safety arising from elevated nitrate and nitrite contents of ingested foods.
3. Impacts of animal waste (manure) and human waste (septic and sewage) entering the landscape.

Nitrogen fertilizer not only increases cereal crop yields, but it also typically improves grain protein concentration (Blumenthal et al. 2001). By contrast, recent reductions of N fertilizer inputs have resulted in reduced grain quality for wheat in Northern Europe (Knudsen 2003). Increasing crop yields may also indirectly improve human health by enhancing household income through the sale of inputs (i.e., small-scale fertilizer sales) and excess produce.

One long-held concern is that ingesting nitrate-rich food and drinking water may be harmful to human health (Townsend et al. 2003). The intake of nitrate in vegetables accounts for more than 80 percent of the nitrate ingested by humans in the United States and 60 percent in the UK, whereas drinking water usually provides only a minor portion (2–25 percent) of the body's external intake of nitrate (L'hirondel and L'hirondel 2002). Nitrate concentrations in vegetables vary widely according to species, maturity, fertilization, and light intensity, but mean values can reach greater than 2500 mg NO_3 kg^{-1}. The high concentrations of plant nitrate are associated with excessive applications of mineral fertilizers or manure, although the relationship is neither very close nor systematic (Greenwood and Hunt 1986).

Excessive nitrate intake has been linked to various forms of cancer. Although ingested nitrate is not thought to be carcinogenic, some ingested nitrate may be converted to nitrite in the body. Laboratory studies on animals and limited studies of exposure in humans suggested that cancer may be induced by nitrosamines and nitrosamides formed as the result of nitrite reacting with amines and amides (Follett and Follett 2001). Nitrite can also restrict hemoglobin's ability to transport oxygen, and limited studies between the 1940s and 1960s linked high nitrate contents in well water with methemoglobinemia (Follett and Follett 2001). Extensive research has failed, however, to identify nitrate conclusively as the cause of increased risk of cancer, and it has been proposed that microbial conversion of nitrate to nitrite in contaminated well water may have been a dominant factor in earlier studies on methemoglobinemia (L'hirondel and L'hirondel 2002). Indeed, new evidence now points to beneficial effects of dietary nitrate (Addiscott and Benjamin 2004). For example, it has been demonstrated that both nitric oxide (NO) and peroxynitrite ($ONOO^-$) can be formed in the human body from nitrate. Both compounds have an antifungal and antibacterial effect against organisms such as *Salmonella, Escherichia coli,* and *Helicobacter pylori.* Other studies also suggest that nitrate protects against cardiovascular diseases (Jenkinson 2001). The net result of these findings from past and recent research is that currently little consensus exists about the risk to human health from consuming nitrate in food and water.

Summary of Potential Off-site Effects via the Gaseous Phase

Of the four major combined N gases released into the atmosphere as a consequence of human activities (NO, NO_2, N_2O, and NH_3), agriculture is believed to be a major source of two of them, NH_3 and N_2O (Jenkinson 2001). Galloway et al. (1995) estimated that more than two thirds of the NH_3-N produced globally each year as a result of human activity was either associated with domestic animals or was volatilized from fertilized fields. Direct emissions of N_2O from agricultural soils are in the order of two to three million tons of N_2O-N (Table 4.2), and estimates of indirect emissions from N after it is leached or eroded from the site of application suggest that this may be of similar magnitude (Mosier 2001). Further losses of N_2O can also be expected to result from waste management of livestock excreta (Mosier 2001).

The health and environmental implications associated with gaseous forms of N include the following:

1. Respiratory and cardiac disease induced by exposure to high concentrations of ozone produced by reactions of NO/N_2O with oxygen in the atmosphere and fine particulate matter.
2. Reactive nitrogen oxides and NH_3 interact with substances in the atmosphere to create hydroscopic aerosols that can act as condensation nuclei for clouds. The increase in atmospheric aerosols, besides their contribution to acid deposition, causes some climate feedbacks and regional problems such as decreased visibility.
3. Depletion of stratospheric ozone by N_2O emissions.
4. Global climate change induced by emissions of N_2O and formation of tropospheric ozone.
5. Ozone-induced injury to crop, forest, and natural ecosystems and predisposition to attack by pathogens and insects.
6. Increased productivity of N-limited natural ecosystems following N deposition and N saturation of soils in forests and other natural ecosystems.
7. Following deposition in rainfall, acidification and eutrophication effects on forests and soils, biodiversity changes in terrestrial ecosystems, and invasion by "weedy" species.

Nitrous oxide is a potent greenhouse gas, with a long half-life in the atmosphere (110 to 150 years, Peoples et al. 1995). Emissions of NO and NO_2 into the atmosphere contribute to acid deposition, which loads atmospheric acid to the ecosystems in the forms of gases, particles, and liquid. Both lead to air pollution through the production of other photochemical oxidant species in the atmosphere, such as ozone.

Because NH_3 has a short life in the atmosphere, it can provide a secondary source for the formation of NO and N_2O (Peoples et al. 1995). The deposition of NH_3 and its ionized form, NH_4^+ also has acidification potential because it is easily transformed to nitrate through nitrification in soils with the concomitant production of protons (Kennedy 1992). Acid deposition has particular significance for natural terrestrial and aquatic ecosystems.

Summary of Potential Off-site Effects Via the Liquid Phase

The following are impacts of fertilizer N lost by runoff, erosion, and leaching:

1. Nitrate and nitrite contamination of ground and surface waters
2. Acidification of freshwater aquatic ecosystems
3. Blooms of toxic algae, with potential harmful effects to humans and animals
4. Eutrophication and hypoxia in coastal ecosystems
5. Possible increase in the incidence of human diseases, such as cholera outbreaks associated with coastal algal blooms
6. Supposed increases in disease vectors such as mosquito hosts of malaria and West Nile virus as a result of increased concentrations of inorganic N in surface water
7. Biodiversity shifts in aquatic ecosystems

Many of the world's surface and ground waters are degraded from nutrient pollution. One of the most serious and well-studied effects of nutrient loadings to waters is that of eutrophication. Nitrogen plays a major role, especially in estuaries, where it is typically the limiting nutrient (Vitousek et al. 1997). Coastal eutrophication is thought to be one of the most widespread pollution problems in the world (Howarth et al. 1996). Eutrophication may result in interrelated consequences, such as rapid growth of blue-green algae and macrophytes, depletion of oxygen in surface water (hypoxia), disappearance of aquatic biodiversity, and production of toxins that are poisonous to fish, cattle, and humans (Rabalias 2002). Many studies have shown that there is a direct and positive correlation between total net N inputs to landscapes and riverine N export (e.g., Van Drecht et al. 2003). Major drivers behind the N increase in surface waters are the increasing N inputs to landscapes from population growth, agricultural intensification, and atmospheric N deposition from fossil fuel combustion (Howarth et al. 1996), with agricultural N sources playing a major role (Boyer et al., Chapter 16, this volume).

Conclusions

The use of N fertilizers has greatly increased global food production, aiming to benefit society by satisfying the needs and demands of a growing world population. Although the benefits are numerous, there are associated costs in the form of environmental impacts, largely associated with losses of managed N inputs to air and water. Some evidence of smaller losses from fertilizers than manures has been found. As fertilizer use increases, losses are also likely to increase. Possibilities of setting targets to minimize losses and maximize NUE are detailed elsewhere (Balasubramanian et al., Chapter 2, this volume). The appropriate use of N will not only reduce the risk of N losses and undesirable consequences, but it will also optimize the plant's ability to utilize other nutrients and scarce resources such as water, improve the ability to manage soil degradation, and reduce the pressure to expand agricultural land into marginal areas and the loss of native habitat.

The positive benefits of N fertilizers for better food quantity and quality are well proven, as are some of the negative impacts of N on the environment; however, evidence for some of the presumed negative impacts on human health remains inconclusive. Making informed judgments about priorities for remedying any of the negative health or environmental impacts of N use requires more than filling gaps in scientific knowledge. The design of strategies to mitigate against problems arising from N supply requires a balanced assessment of both the potential benefits and costs arising from such losses as well as of the costs of mitigation. In practically all cases, it is not yet possible to make such assessments.

The need to identify the relative contribution of fertilizer N to environmental and health problems on a regional basis is ongoing. In some regions, inputs other than fertilizers may be the main polluter, such as atmospheric deposition or animal and human waste (Howarth et al. 1996). Synthetic fertilizer N inputs are a major source of reactive N inputs to terrestrial landscapes on a global scale, representing more than 60 percent of the total net anthropogenic N inputs to the terrestrial landscape in the 1990s (Boyer et al., Chapter 16, this volume). The direct relationship between N inputs to landscapes and N exports to the coastal zone, where some of the most widespread consequences of N losses are manifested in the forms of eutrophication, underscores the need to seek strategies to minimize N losses from agricultural systems. More research is needed on the fate of the various storage pools and loss processes, especially how much N is lost as N_2 during denitrification. This could well help to close the gap of "unaccounted for N" in budgets.

Literature Cited

Addiscott, T. M., and N. Benjamin. 2004. Nitrate and human health. *Soil use and management* 20:98–104.

Becker, M., and J. K. Ladha, and J. C. G. Ottow. 1994. Nitrogen losses and lowland rice yield as affected by residue N release. *Soil Science Society of America Journal* 58:1660–1665.

Berntsen, J., B. H. Jacobsen, J. E. Olesen, B. M. Petersen, and N. J. Hutchings. 2003. Evaluating nitrogen taxation scenarios using the dynamic whole farm simulation model FASSET. *Agricultural Systems* 76:817–839.

Blumenthal J. M., D. D. Baltensperger, K. G. Cassman, S. C. Mason, and A. D. Pavlista. 2001. Importance and effect of nitrogen on crop quality and health. Pp. 45–63 in

Nitrogen in the environment: Sources, problems, and management, edited by R. F. Follett and J. L. Hatfield. Amsterdam: Elsevier Science.

Børgesen, C. D., J. Djurhuus, and A. Kyllingsbæk. 2001. Estimating the effect of legislation on nitrate leaching by upscaling field simulations. *Ecological Modelling* 136:31–48.

Brown, L.R. 1999. Feeding nine billion. Pp.115–132 in *State of the world,* edited by L. R. Brown, C. Flavin, and H. French. New York: W. Norton & Co.

Crews, T. E., and M. B. Peoples. 2004. Legume versus fertilizer sources of nitrogen: Ecological tradeoffs and human needs. *Agriculture, Ecosystems and Environment* 102:279–297.

Diekmann, F. H., S. K. DeDatta, and J. C. G. Ottow. 1993. Nitrogen uptake and recovery from urea and green manure in lowland rice measured by ^{15}N and non-isotope techniques. *Plant and Soil* 147:91–99.

Fillery, I. R. P. 2001. The fate of biologically fixed nitrogen in legume-based dryland farming systems: A review. *Australian Journal of Experimental Agriculture* 41:361–381.

Follett J. R., and R. F. Follett. 2001. Utilization and metabolism of nitrogen by humans. Pp. 65–92 in *Nitrogen in the environment: Sources, problems, and management,* edited by R. F. Follett and J. L. Hatfield. Amsterdam: Elsevier Science.

Galloway, J. N., and E. B. Cowling. 2002. Reactive nitrogen and the world: 200 years of change. *Ambio* 31:64–71.

Galloway, J. N., W. H. Schlesinger, H .Levy, A. Michaels, and J. L. Schnoor. 1995. Nitrogen fixation: Atmospheric enhancement—environmental response. *Global Biochemical Cycles* 9:235–252.

Gök, M., and J. C. G. Ottow 1986. Effect of cellulose and straw incorporation on soil denitrification and nitrogen immobilization at initially aerobic and permanent anaerobic conditions. *Biology and Fertility of Soils* 5:317–322.

Greenwood, D. J., and J. Hunt. 1986. Effect of nitrogen fertiliser on the nitrate contents of field vegetables grown in Britain. *Journal of the Science of Food and Agriculture Cambridge* 37:373–383.

Hansen, S., H. E. Jensen, N. E. Nielsen, and H. Svendsen. 1991. Simulation of nitrogen dynamics and biomass production in winter wheat using the Danish simulation model DAISY. *Fertilizer Research* 27:245–259.

Harris, G. H., O. B. Hesterman, E. A. Paul, S. E. Peters, and R. R. Janke. 1994. Fate of legume and fertilizer nitrogen-15 in a long-term cropping systems experiment. *Agronomy Journal* 86:910–915.

Howarth, R. W., G. Billen, D. Swaney, A. Townsend, N. Jaworski, K. Lajtha, J. A. Downing, R. Elmgren, N. Caraco, T. Jordan, F. Berendse, J. Freney, V. Kudeyarov, P. Murdoch, and Z. Zhu. 1996. Regional nitrogen budgets and riverine N & P fluxes for the drainages to the North Atlantic Ocean: Natural and human influences. *Biogeochemistry* 35:75–139.

IFA/FAO. 2001. *Global estimates of gaseous emissions of NH_3, NO and N_2O from agricultural land.* Rome: FAO.

Janzen, H. H., J. B. Bole, V. O. Biederbeck, and A. E. Slinkard. 1990. Fate of N applied as green manure or ammonium fertilizer to soil subsequently cropped with spring wheat at three sites in western Canada. *Canadian Journal of Soil Science* 70:313–323.

Jenkinson, D. S. 2001. The impact of humans on the nitrogen cycle, with focus on temperate agriculture. *Plant and Soil* 228:3–15.

Kennedy, I. R. 1992. *Acid soil and acid rain.* New York: John Wiley and Sons.

Knudsen, L. 2003. *Nitrogen input controls on Danish farms: Agronomic, economic and environmental effects.* Proceedings No. 520, November 2003. UK: International Fertiliser Society. (www.fertiliser-society.org/Content/Publications.asp)

Kumar, K., and K. M. Goh. 2000. Crop residues and management practices: Effects on soil quality, soil nitrogen dynamics, crop yield, and nitrogen recovery. *Advances in Agronomy* 68:197–319.

Ladd, J. N., and M. Amato. 1986. The fate of nitrogen from legume and fertilizer sources in soils successively cropped with wheat under field conditions. *Soil Biology and Biochemistry* 18:417–425.

Ledgard, S. F. 2001. Nitrogen cycling in low input legume-based agriculture, with emphasis on legume/grass pastures. *Plant and Soil* 228:43–59.

L'hirondel, J., and J. L. L'hirondel. 2002. *Nitrate and man: Toxic, harmless or beneficial?* Wallingford, U.K: CABI Publishing.

Mosier, A. 2001. Exchange of gaseous nitrogen compounds between agricultural systems and the atmosphere. *Plant and Soil* 228:17–27.

Nziguheba, G., R. Merckx, C. A. Palm, and M. R. Rao. 2000. Organic residues affect phosphorus availability and maize yields in a Nitisol of western Kenya. *Biology and Fertility of Soils* 32:328–339.

Parton, W. J., E. A. Holland, S. J. Del Grosso, M. D. Hartman, R. E. Martin, A. R. Mosier, D. S. Ojima, and D. S. Schimel. 2001. Generalized model for NO_x and N_2O emissions from soils. *Journal of Geophysical Research* 106:17403–17419.

Peoples, M. B. 2002. Biological nitrogen fixation, contributions to agriculture. Pp. 103–106 in *Encyclopedia of soil science*, edited by R. Lal. New York: Marcel Dekker.

Peoples, M. B., J. R. Freney, and A. R. Mosier. 1995. Minimizing gaseous losses of nitrogen. Pp. 565–602 in *Nitrogen fertilization and the environment*, edited by P. E. Bacon. New York: Marcel Dekker.

Rabalias, N. N. 2002. Nitrogen in aquatic ecosystems. *Ambio* 31:102–112.

Rochester, I. J., M. B. Peoples, N. R. Hulugalle, R. R. Gault, and G. A. Constable. 2001. Using legumes to enhance nitrogen fertility and improve soil condition in cotton cropping systems. *Field Crops Research* 70:27–41.

Sisti, C. P. J., H. P. dos Santos, R. Kohhann, B. J. R. Alves, S. Urquiaga, and R. M. Boddey. 2004. Change in carbon and nitrogen stocks in soil under 13 years of conventional or zero tillage in southern Brazil. *Soil and Tillage Research* 76:39–58.

Smil, V. 2001. *Enriching the Earth.* Cambridge, Mass: MIT Press.

Smith, P., J. U. Smith, D. S. Powlson, W. B. McGill, J. R. M. Arah, O. G. Cheroot, K. Coleman, U. Franco, S. Frolking, D. S. Jenkinson, L. S. Jensen, R. H. Kelly, H. Klein Gunnewiek, A. S. Komarov, C. Li, J. A .E. Molina, T. Mueller, W. J. Parton, J. H. M. Thornley, and A. P. Whitmore. 1997. A comparison of the performance of nine soil organic matter models using datasets from seven long-term experiments. *Geoderma* 81:153–225.

Sommer, S. G., S. Géneremont, P. Cellier, N. J. Hutchings, J. E. Olesen, and T. Morvan. 2003. Processes controlling ammonia emission from livestock slurry in the field. *European Journal of Agronomy* 19:465–486.

Townsend, A. R., R. W. Howarth, F. A. Bazzaz, M. S. Booth, C. C. Cleveland, S. K. Collinge, A. P. Dobson, P. R. Espstein, E. A. Holland, D. R. Keeney, M. A. Mallin, C. A. Rogers, P. Wayne, and A. H. Wolfe. 2003. Human health effects of a changing global nitrogen cycle. *Frontiers in Ecology and the Environment* 1:240–246.

van Cleemput, O. 1998. Subsoils: Chemo- and biological denitrification, N_2O and N_2 emissions. *Nutrient Cycling in Agroecosystems* 52:187–194.

Van Drecht, G., A. F. Bouwman, J. M. Knoop, A. H. W. Beusen, and C. R. Meinardi. 2003. Global modeling of the fate of nitrogen from point and nonpoint sources in soils, groundwater, and surface water. *Global Biogeochemical Cycles* 17,1115:26.1–26.20.

Vanlauwe, B., N. Sanginga, and R. Merckx. 1998. Recovery of *Leucaena* and *Dactyladenia* residue ^{15}N in alley cropping systems. *Soil Science Society of America Journal* 62:454–460.

Vanlauwe, B., N. Sanginga, and R. Merckx. 2001a. Alley cropping with *Senna siamea* in Southwestern Nigeria: I. Recovery of N-15 labeled urea by the alley cropping system. *Plant and Soil* 231:187–199.

Vanlauwe, B., J. Wendt, and J. Diels. 2001b. Combined application of organic matter and fertilizer. Pp. 247-280 in *Sustaining soil fertility in West-Africa*, edited by G. Tian, F. Ishida, and J. D. H. Keatinge. SSSA Special Publication No. 58. Madison, Wisconsin: Soil Science Society of America.

Vanlauwe, B., J. Diels, K. Aihou, E. N. O. Iwuafor, O. Lyasse, N. Sanginga, and R. Merckx. 2002. Direct interactions between N fertilizer and organic matter: Evidence from trials with ^{15}N labelled fertilizer. Pp. 173–184 in *Integrated plant nutrient management in sub-Saharan Africa: From concept to practice*, edited by B. Vanlauwe, J. Diels, N. Sanginga and R. Merckx. Wallingford, UK: CABI Publishing.

Vitousek, P. M., J. D. Aber, R. W. Howarth, G. E. Likens, P. A. Matson, D. W. Schindler, W. H. Schlesinger, and D. G. Tilman. 1997. Human alteration of the global nitrogen cycle: Sources and consequences. *Ecological Applications* 7:737–750.

5

Societal Responses for Addressing Nitrogen Fertilizer Needs: Balancing Food Production and Environmental Concerns

Cheryl A. Palm, Pedro L. O. A. Machado, Tariq Mahmood, Jerry Melillo, Scott T. Murrell, Justice Nyamangara, Mary Scholes, Elsje Sisworo, Jørgen E. Olesen, John Pender, John Stewart, and James N. Galloway

A basic need of any society is an affordable, secure, high-quality food supply. The ability of the agricultural sector to fulfill this need determines whether a society can support itself or whether it must rely on food from other areas. Optimum agricultural production depends on an adequate supply chain as well as an effective distribution and marketing system, and as such it depends on the effectiveness with which the agricultural sector interacts with other sectors and how this is influenced by agricultural and environmental policies (Mosier et al., Chapter 1, this volume).

Nitrogen (N) is the major nutrient required for crop production because it is often deficient in agricultural soils and is an essential component of proteins. The nature of N fertilization issues vary across the globe. Situations of insufficient N application with resulting food insecurity and environmental degradation describe much of sub-Saharan Africa (SSA) and many parts of Central America; the other extreme of excess application and N pollution is found in Western Europe, the United States, and more recently China. Ultimately, there is need to balance the benefits derived from N applications with the associated environmental costs.

In this chapter we explore the societal priorities and policies that lead to excess or inadequate application of N or that can help mitigate the associated environment and health effects. Our approach was to classify N use broadly, based on supply/access to N fertilizers and N application rates (Figure 5.1). The various categories of N supply and use come at a variety of scales. Although supply at the national scale should generally

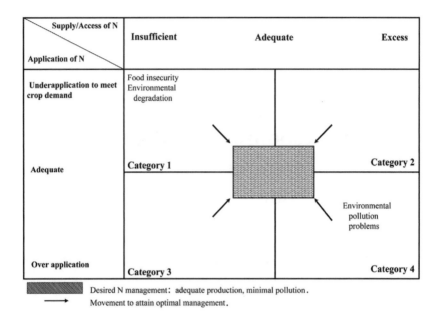

Figure 5.1. Matrix showing the range of N access/supply and N application rates that emerge during the development of agriculture and the consequent effects on food security and the environment.

meet demand, some developing countries still have an insufficient N supply, whereas others have a sufficient supply at the national scale but little or no access to that supply in some parts of the country or society.

The transition from inadequate N application to excessive N application characterizes the agricultural development pathway in most societies. Achieving food security is the concern early in the development process with little attention or ability to redress environmental degradation or pollution issues. As the agriculture and economy of the country develop, there comes a point of ample, affordable, high-quality food resulting from adequate or even excessive N applications. At some stage the concerns of society and governments may shift to the environmental impacts of agricultural activities.

The goal for N management, indicated by the arrows and box in Figure 5.1, should be N access, application, and management that result in adequate and sustainable food production that contributes to economic growth and minimizes environmental pollution. The route to this goal would preferably be from insufficient supply and application (category 1) directly to adequate supply and management of N for food production and the environment (category 2) while avoiding situations of excess supply and application (category 4) altogether.

In the following section, a set of case studies illustrates various situations in the N supply/N application matrix (Figure 5.1 and Table 5.1).

Table 5.1. Characterization of the case studies

Location	Matrix position	Climatic zone	Farm size (ha)	Application rate (kg N ha^{-1} yr^{-1})	Source of applied N	Major problems
Sub-Saharan Africa Smallholders (Zimbabwe)	Category 1 Insufficient supply Insufficient application	Semi-arid	0.1–1	5 30	Fertilizer Manure	Nutrient depletion Soil erosion Food insecurity
South Asia (Pakistan)	Category 2 Adequate supply Insufficient application	Semi-arid Subtropical	4.4	102	Fertilizer	NH$_3$ volatilization Denitrification Food insecurity
Planted grass pasture (Brazil)	Category 2 Adequate supply Insufficient application	Tropical Subtropical	50–3700	insignificant	—	Soil degradation Deforestation
Asia (Indonesia, Philippines, China)	Categories 2–4 Adequate supply Adequate to excess application	Tropical	1	100–300 (season)	Fertilizer	NH$_3$ emission Surface-water pollution
North America (Midwest USA)	Category 4 Adequate supply Excess application	Temperate	238	150	Fertilizer	NO$_3$ leaching NO$_x$ emission Surface runoff
NW Europe (Netherlands, Denmark)	Category 4 Adequate supply Excess application	Humid Temperate	30–50	200 (DK) 500 (NL)	Fertilizer Manure	Eutrophication of land/water Groundwater contamination

Case Studies

Inadequate Supply and Insufficient Application (Category 1)

ZIMBABWE: NUTRIENT DEPLETION ON SMALLHOLDER FARMS

Declining soil fertility is a major constraint to sustainable smallholder crop production in SSA, leading to environmental degradation, yield decreases, and food insecurity (Sanchez 2002). The smallholder farm in Zimbabwe is characterized by small farm size (<1 ha per household), infertile sandy soils, continuous cultivation, and suboptimal inputs.

Low fertilizer application rates are widespread in SSA, where average fertilizer use in 1999 was only 2.8 kg of NPK fertilizer per ha of agricultural land (8.4 for Zimbabwe) (FAOSTAT 2004). In 1989 through 1990, when fertilizers were still subsidized in Zimbabwe, smallholder farmers applied 53 kg ha^{-1} NPKS fertilizer compared with 705 kg ha^{-1} by large-scale commercial farmers (Humphreys 1991). This indicates inequities in access (category 3) but also abuse of subsidies in the commercial sector. Fertilizer subsidies in Zimbabwe are currently extremely limited. Annual soil N depletion resulting from continuous cultivation and limited fertilizer use in Zimbabwe was estimated to exceed 20 kg N ha^{-1} (Smaling et al. 1997). Fertilizer use by Zimbabwean smallholder farmers is most constrained by the availability of cash (FAO 1999). The proportion of household income allocated to fertilizer purchases varies between 16 and 62 percent, where even at the higher percentage the fertilizer applied is still suboptimal.

Investment in physical and social infrastructure in smallholder areas has lagged, and smallholder farmers receive fertilizer and other inputs too late in the planting season because of the poor condition of roads, shortage of transport vehicles, and low profitability of servicing smallholder farmers' needs. The farmers pay higher farm-gate prices for fertilizers and sell their produce at uneconomic prices because of poor marketing infrastructure and lack of competition among suppliers (FAO 1999).

Three quarters of smallholder farmers are aware of the benefits of fertilizer application but have limited knowledge of appropriate management practices (FAO 1999); nor do extension officers based in smallholder farming areas have transport to visit the large number of farmers. Other constraining factors include lack of access to credit and markets and unreliable rainfall.

Farmers, communities, and governmental departments have responded in a wide range of ways to overcome this insufficient application of N. In many areas, farmer groups have formed to enable them to purchase fertilizer directly from manufacturers at lower prices and also to bargain for lower transport costs. These groups are also used for marketing produce and are also focal points for extension officers to disseminate new technologies to the farmers at reduced cost. Manufacturers have introduced smaller fertilizer packs to enable poor farmers to afford some fertilizer. Fertilizer formulas have been

Table 5.2. Approaches that would improve access to application of nitrogen and reverse soil degradation and food security in the smallholder sector of Zimbabwe

Farmer constraints	Level of societal responses	Policy tools	Limitations
Lack of cash and credit	Formation of farmer cooperatives; repacking of fertilizers for easier distribution by cooperatives	Enabling environment for financial institutions to extend credit to farmers	Lack of collateral
Poor infrastructure and transport	Community- and national-based strategy for the rehabilitation of roads	Existing district Development Funds to improve existing roads	Funds, equipment and mismanagement
Limited knowledge on fertilizer management	Limited training and consultations by cooperatives, farmers unions, agro-dealers and universities	Strengthening of the extension networks and farmer training	Funds
Unreliable rainfall	Small-scale irrigation and water-harvesting projects	Strategies for water conservation and enhanced weather forecasting capabilities	Funds and equipment

changed to increase nutrient concentrations, thereby reducing transport costs. The government has maintained subsidies for ammonium nitrate to promote the application of N, the most limiting nutrient in the country. These responses have led to only limited improvement in food security and soil degradation in Zimbabwe. The need to implement a variety of additional responses, including access to credit, rehabilitation of roads, repacking fertilizers into "affordable" quantities, training, and small-scale irrigation and water harvesting projects is urgent (Table 5.2).

Adequate Supply, Inadequate Application (Category 3)

Brazil: Planted Pasture

Planted pastures in Brazil receive little to no N fertilizer despite the availability of appropriate technologies and sufficient supplies. Pasture in Brazil covers 178 M ha, approximately 105 M ha of which are planted pastures (Zimmer and Euclides-Filho 1997). Meat production is based on extensive grazing systems with no fodder or supplementation apart from mineral salt. Pasture grasses are planted 1 or 2 years after clearing and burning of natural vegetation (savanna, Atlantic, or humid forest) and arable cropping. No fertilizer is applied, and pasture plants benefit either from the ashes from burning or residual lime and P, K fertilizer applied to arable crops. Pasture productivity declines after 2 to 5 years, nonpalatable weeds invade, and patches of bare soil appear. Degraded pasture areas are abandoned, and new areas of natural vegetation are converted to pasture. Degrading pastures are now estimated at 50 M ha or half the planted pasture area (Kichel and Miranda 1999). This situation is caused by low land prices resulting from abundant area of lands in natural vegetation, governmental policies for regional development, and increasing national and international markets for beef.

Technologies exist for improved pasture management and increased beef production, including the use of forage legumes and mineral N fertilizers (Oliveira et al. 2001), but in Brazil ranchers have made minimal adaptation to impede or reclaim pasture degradation. The constraints for moving the production system to sustainable pasture production are the lack of incentives by the government and inadequate education or extension services. To reverse this situation, it is necessary to determine whether ranchers are carrying out this degradation knowingly or not. This will help representatives of the national sector (e.g., research scientists from state institutes) or from nongovernmental organizations (NGOs, e.g., cattle-breeders associations) to implement effective measures.

Although Brazil has passed legislation aimed at reducing deforestation, implementation suffers from insufficient funding for policing and enforcement. The need for trained federal or state agents to enforce these laws is urgent, and a mechanism to penalize unauthorized clearing of natural vegetation is needed. To be most effective, simultaneous educational programs could demonstrate to ranchers how pasture systems can be improved and economically sustained.

PAKISTAN: SMALLHOLDER CROP PRODUCTION

The agricultural growth rate of 4.2 percent in Pakistan is far below its potential. Average farm size is 4.4 ha; 81 percent of the farms are 0.5 to 5 ha (smallholdings), 7 percent are 10 to 20 ha, and only 0.3 percent are greater than 60 ha and include progressive farmers with sufficient financial resources. Many of the problems occur in the smallholder sector, resulting in suboptimal use of N fertilizer, averaging 103 kg N ha^{-1} yr^{-1} compared with 240 kg N ha^{-1} year^{-1} applied by growers on larger land holdings. Crop yields in the smallholder sector are one third those obtained by farmers who can afford adequate fertilizer and other inputs. Although Pakistan is self-sufficient in major agricultural commodities, occasionally some commodities must be imported to meet national requirements, and food insecurity and malnutrition are often more pronounced among smallholders. Reasons for underapplication in the smallholder sector are the relatively high cost of unsubsidized fertilizers and other inputs and the inefficiency of small-scale systems. Although little information is available, the efficiency of nitrogen use (NUE) appears to be low and N losses high, with some ^{15}N-balance studies indicating total fertilizer N losses from 33 to 42 percent (Mahmood et al. 1998).

Economic constraints to the application of adequate N and improved management include limited access and timely distribution of agricultural credit and the high cost of agricultural inputs leading to high production costs. Agronomic constraints include nutrient imbalances in fertilizer applications and inadequate water supply. Infrastructure constraints include inefficient irrigation systems that can lead to more than 50 percent water loss, limited number of soil testing laboratories, and the poor condition of roads to the farm gate.

Adequate supply of agricultural credit and price support for fertilizers or crops might be necessary to overcome financial constraints and to encourage increased fertilizer applications. Subsidized electricity and fuel in the agricultural sector could play a vital role in alleviating financial constraints because supplementary irrigation with groundwater is very cost intensive. Improving road and irrigation infrastructure would increase production and market access that in turn would enable a more efficient system to develop. Increased research and extension would assist in the development of more efficient cropping systems, including the distribution and use of balanced fertilizers.

Adequate/Excessive Supply Plus Overapplication (Category 4)

WESTERN EUROPE: INTENSIVE LIVESTOCK AND CROP FARMING SYSTEMS

Losses of N from European agricultural activities have caused considerable environmental concerns. Eutrophication of terrestrial, aquatic, and coastal ecosystems has increased public awareness and concern. Additionally, nitrate contents in drinking water in some areas exceed the threshold value of 50 mg NO_3 l^{-1} noted in the European Union (EU) "Nitrate Directive." These problems are often found in areas with high live-stock densities, where both animal manures and mineral fertilizer N have been applied indiscriminately to agricultural soils.

This case is illustrated for Denmark and The Netherlands, both of which have high livestock densities (Table 5.3). Agricultural activities in both countries have responded to the EU Common Agricultural Policy (CAP), where price supports have stimulated intensification in agricultural production. The N output is higher per land area in The Netherlands as a result of a higher proportion of grassland and a high water table, result-ing in denitrification as the dominant loss pathway. In Denmark nitrate leaching dom-inates as the loss pathway in cereal cropping systems on the freely drained soils.

Reductions in N losses are constrained by economic losses from reduced livestock and crop production and costs for improved manure management. Additional costs are associated with the development and implementation of new technologies. Political will must be strong to develop and enforce new and existing environmental legislation that implies considerable costs for both the private and public sectors.

In the 1990s, the EU addressed environmental side effects of agriculture with the reform of the CAP and various environmental regulations and directives. One result was the "Nitrate Directive," with a goal of reducing nitrates in drainage water to less than 50 mg NO_3 l^{-1}. It requires implementation of mandatory measures related to (1) peri-ods when the application of manure and fertilizers is prohibited; (2) capacity and facil-ities for storing animal manure; and (3) limits to the amounts of manure and fertiliz-ers applied. There is wide variation among members in the implementation of this Directive (De Clercq et al. 2001).

Regulation in The Netherlands has been implemented through the Mineral Accounting System (MINAS), a farm-gate balance of all N and P inputs and outputs. The MINAS system applies a levy for the surplus nutrients that exceed critical thresh-olds depending on livestock density. The costs of the levies vary with farm type (1000 to 5000 euros per farm per year) (Oenema 2004). Denmark has implemented a more rigid framework for reducing N inputs (Table 5.4; Olesen et al., Chapter 9, this vol-ume) through mandatory regulations with no financial compensation to the farmers. Costs associated with these regulations have been estimated to be 100 to 200 million euros per year. Some voluntary and subsidized schemes have also been implemented (Table 5.4). In both countries, research, demonstration, and extension have been cru-cial for achieving cost-effective implementation of reduced nutrient loading to the environment.

The measures have reduced N surplus by 32 and 38 percent in The Netherlands and Denmark, respectively, from 1995 to 2003. This was accomplished partially by reduc-ing livestock density in The Netherlands, costing 0.5 billion € to reduce the number of pigs by 10 percent. In Denmark, livestock production increased slightly, reflecting an increase in NUE. The regulations in Denmark were approved by the EU Commission,

Table 5.3. Basic agricultural statistics and nitrogen input and output to agricultural soils in The Netherlands and Denmark in 1995

Category	Item	The Netherlands	Denmark
Agricultural statistics	Agricultural area (M ha)	2.0	2.7
	Livestock density (LU ha^{-1})	3.9	1.6
	Cereal area (%)	11	56
	Grassland area (%)	53	14
Nitrogen input (Gg yr^{-1})	Fertilizer	406	311
	Manure	630	231
	Deposition (NH$_4$ + NO$_x$)	76	44
	Other	37	62
Nitrogen output (Gg yr^{-1})	Yield of crops and grass	448	317
	Ammonia volatilization	146	48
	Leaching and runoff	86	235
	Denitrification + accumulation	469	48
Nitrogen surplus (Gg yr^{-1})	All inputs, yield	701	306
	Fertilizers + manure, yield	588	200
N recovery efficiency (RE)	Yield over all inputs	39	51
	Yield over fertilizers + manure	43	61

From Kroeze et al. 2003; Kyllingsbæk 2000; FAOSTAT 2004.

Table 5.4. Measures applied in Denmark to reduce nitrate leaching in the Aquatic Action Plans.

Type	Measure
Mandatory	Requirements for 65% winter-green crops
	Additional 6% cover crops (on top of the winter green crops)
	Restrictions on livestock densities (depend on livestock type)
	Increasing requirements for utilization of N in manure (depend on manure type)
	Maximum rates of N application (10% below economical optimum)
Optional (subsidized)	Organic farming
	Conversion of farmland to permanent wetlands
	Afforestation on agricultural land
	Agreements with farmers on environmentally friendly management

but those of The Netherlands were not, implying that a more restrictive system, similar to that in Denmark, may be required.

Future regulation of N use in EU countries will be dominated by the "Water Framework Directive," which requires all inland and coastal waters to reach good ecologic status by 2015. This will be done by establishing a river basin district within which environmental objectives will be set. Regulations may differ by water catchments, depending on the impact on the protected water body.

THE U.S. MIDWEST: CORN PRODUCTION SYSTEMS

Overapplication of N in corn production systems in the U.S. Midwest arise (1) where farmers apply extra N to ensure against yield-limiting N losses, (2) from generalized N recommendations not tailored to local conditions, (3) from poor growing season conditions, (4) from unrealistic yield goals, (5) where soil N release is greater than expected, (6) where there is poor integration of fertilizer and manure N sources, (7) where other limiting factors suppress crop response, and (8) when applications are not synchronized with crop demand. Although there are several important sources of N input to rivers and coastal systems, N fertilizer use has been shown to parallel trends in riverine flux of N to the coast (Howarth et al. 2002).

Constraints to improved N management include the low price of N fertilizer relative to crop price, making the financial risk of underapplications greater than that of overapplications. Downsizing agricultural research and education programs also is an impediment to developing and communicating improved N management. Even with the availability of some improved N management practices, the perceptions of acceptable risk by farmers affect their decisions in adopting them. Many farmers and input suppliers are unwilling to accept certain practices, such as spring application of N, because of the strain on their resources during the planting period.

Improved N management could be approached through education, technical assistance, and incentive programs. Since 1985, farm income support programs have been tied to soil conservation and improved management on highly erodable lands as well as to the preservation of wetlands (Claassen et al. 2001). Important to N management is that these programs have reduced soil erosion, created more wetlands than have been lost, and increased carbon sequestration (Claassen et al. 2001). Programs also exist that encourage conservation practices on productive farmlands. The Environmental Quality Incentives Program (EQIP) provides cost sharing for the implementation of conservation practices, and the Conservation Security Program (CSP) provides funding to producers who have already incorporated such management (Claassen 2003). Both EQIP and CSP consider improved nutrient management, but EQIP has substantially more funding to do so.

In the United States, some regulatory programs target fertilizer N applications on lands receiving manure from some animal feeding operations considered point sources of pollution under the Clean Water Act (U.S. Department of Agriculture and the Environmental Protection Agency 1999). For coastal states, an additional regulatory program, the Coastal Zone Act Reauthorization Amendments, calls for reduction of non–point source pollution into coastal waters.

Food labeling and insurance programs are other interventions that could lead to improved N management. Food labeling that reflects environmentally sound management practices allows the public to make more informed choices about the kinds of production practices it wants to endorse through its food purchases. A new insurance program is being offered to farmers for following best management practices (BMP) (Minnesota Department of Agriculture 2003). This program provides payment if the implementation of BMPs results in yield reductions of more than 5 percent compared with conventional practices.

Asia: Intensive Rice Production—Systems in Transition

Rice is cultivated extensively in Asia with the widespread use of mineral fertilizers. The following three cases represent a transition from adequate to excessive N application. Recommended rates for rice production in the region vary from 90 to 200 kg N ha^{-1}. In economically well-developed regions in China, farmers in the major rice growing areas apply excess N fertilizer, 200–300 kg N ha^{-1} for each crop (Buresh et al., Chapter 10, this volume), resulting in average yields of 6.3 t ha^{-1}. Good infrastructure and a readily available and relatively inexpensive N fertilizer facilitate application and promote overuse in China. In addition, previous government policies in China provided blanket fertilizer recommendations for maximizing yields. Urea and ammonium bicarbonate are the most commonly used fertilizers, and nitrogen losses through ammonia volatilization, leaching, and runoff range from 20 to 70 percent (Zhu and Wen 1992). Concerns have arisen about environmental problems related to leaching and runoff from rice fields and possible N deposition from the ammonia volatilization.

In a rice–vegetable cropping system in the lowlands of Ilocos Norte, Philippines, rice is grown in the wet season with rainwater, in rotation with diverse upland vegetable crops in the dry season with partial or full irrigation. Farmers apply N (as urea) up to 120 kg ha^{-1} with rice and up to 600 kg ha^{-1} with vegetables. N losses ranging from 34 to 549 kg ha^{-1} in the different rice–vegetable systems have been measured (Tripathi et al. 1997). The major pathway of N loss is nitrate leaching.

In intensive irrigated rice areas of West Java, Indonesia, rice production increased dramatically from 1975 to 1990 as a result of the adoption of improved varieties, expansion of land in rice, and an increase in fertilizer application rates. Fertilizers were subsidized in the 1980s, which led to blanket recommendations and high rates of application. With the economic crisis and increase in fertilizer prices in 1997, fertilizer application rates decreased from around 135 kg N ha^{-1} to 100 kg N ha^{-1} (Abdulrachman et al. 2004). Rice yields have remained constant since 1990 at about 4.5 to 5 t ha^{-1}, but population growth continues, soil quality is declining, and agricultural land is being converted to nonagricultural uses. N recoveries of less than 20 percent indicate that there are large N losses and considerable scope for improved N management to increase rice yields without increasing N fertilizer rates.

Research on rice shows that large N losses can be avoided through proper management of fertilizer N, including placement, timing, and rates of application in addition to management of irrigation water. There is, however, inadequate extension and lack of knowledge at the farm level for increasing NUE and decreasing N losses. In addition, excess fertilizer use is mainly dictated by economic returns, indicating that the fertilizer to crop prices still favor overapplication; this is particularly true for vegetable crops, which, in general, bring higher returns. Although the environmental effects of overapplication have been noted in China, this information may not be widely known in the region and hence not perceived as a problem, particularly in countries or areas that are in rather recent transition from food insecurity.

Research and extension services that demonstrate more efficient N management technologies are needed. Leaf color charts and other tools for fine tuning the timing and amounts of N application have been successful elsewhere and should be widely used throughout the region. Farmers should also be provided with information about the environmental and health effects of overapplication in attempts to reverse the growing environmental problems in China and, if possible, to avert such situations from developing elsewhere.

Driving Forces, Constraints, and Societal Responses to Change in Nitrogen Inputs

Several driving forces and constraints to change in N supply and use emerge from the case studies. The ownership and amount of land available for cultivation influence the use of N fertilizer. Smallholders optimize fertilizer rates to maintain and increase production on a fixed area. Where land is more abundant, there may be the opportunity of clearing more natural vegetation because the cost of this action may be less than applying extra fertilizer and other inputs to already available land. Land ownership also strongly influences a farmer's management objectives. Short-term rental agreements often result in unwillingness by farmers to invest in inputs, whereas ownership creates a willingness to make improvements.

The price of fertilizer relative to the market price of agricultural products plays a dominant role in N access and application. When this ratio is low, N fertilizer is relatively inexpensive, making overuse less financially risky than underuse. Conversely, when this ratio is large, N fertilizer becomes unaffordable to many farmers and is underutilized. Prices of both fertilizer and agricultural products are influenced by policies and infrastructure. In developing countries, lack of good infrastructure not only increases the price of mineral N but also makes it less accessible. Government policies that provide fertilizer subsidies or crop and livestock price support systems should result in increased food production and N fertilizer use, but without careful and informed management can lead to excess N use and N pollution problems.

Lack of knowledge about agricultural practices that are agronomically, economically, and environmentally sound can also impede changes in N use. This lack of knowledge can exist in many sectors of society. For instance, lack of awareness of environmental impacts can result in a lack of will by policy makers, communities, advisers, and farmers to improve N management. Such unwillingness for change is, in part, a result of the perception of the existence or degree of risk associated with various practices. Interestingly, unpredictable weather is often indicated as one of the factors that affect N management.

Access to and use of N are closely linked with agricultural and environmental policies. In general, three different policy environments can be distinguished: (1) the neglect of agricultural development leading to land degradation and food insecurities; (2) a domination of agricultural policies that support intensive production systems; and (3) a focus on the multifunctional aspects of agriculture, including the environment, food quality, and safety issues. Often progression in policy instruments occurs. In situations with restricted fertilizer supply (category 1), the most effective policies would be directed toward development of the infrastructure to ensure that fertilizer is accessible at the local scale. When fertilizer is available but not used, the agricultural policy will have to support its use. Subsidizing fertilizer is an effective means toward this end. When fertilizer use increases, food production may be more effectively ensured through product price support. Subsidies and price supports for agricultural production, however, have historically led to intensive production systems causing environmental pollution (category 4). In such situations, agricultural support needs to be dropped or shifted toward supporting land management rather than production volume. At the same time, environmental regulations and associated incentives must ensure that N losses are kept at acceptable levels.

Implicit in the progression from food security to environmental concerns is that agricultural and environmental policies should be linked to minimize the environmental concerns while maintaining agricultural productivity and profitability. The effectiveness of countries in solving the issues of N use in agricultural production are linked ultimately to the political will. The following two examples illustrate the type of political actions that are needed to address the N deficient and N excess situations, respectively.

Insufficient Nitrogen

General neglect of the agricultural sector throughout SSA over the past couple of decades by both national and international institutions has resulted in dramatic decreases in funding for development and agriculture in rural areas. Neglect, combined with market liberalization and trade policies, has led to increases in fertilizer prices relative to commodity prices, cash and credit constraints to purchase inputs, limited access to markets and infrastructure, and limited agricultural technical assistance. Concerted efforts at local, national, and international levels are needed to reverse this trend.

Infrastructure improvement and development for distribution of inputs and goods produced are top priorities for other policies related to food production to be effective. Subsequently, policies to increase mineral fertilizer availability and accessibility become important. In SSA, access to fertilizers, which cost two to six times the world market price (Sanchez 2002), will be possible only when the numerous constraints are removed. Fertilizer subsidies were eliminated in most of SSA in the 1990s. Although there is debate as to whether this actually resulted in decreased fertilizer use, there is now widespread agreement that some type of subsidies must be implemented. More controversial is how to implement these subsidies (i.e., national or local government, NGOs, community-based organizations, the private sector, or some combination and sequence of the different sectors).

Increasing use of fertilizers through subsidies or credit systems will not be effective if prices for crops are not favorable and stable and markets are not developed. Attaining favorable input to crop prices, however, cannot be addressed without considering the enormous impact of the international trade barriers and subsidies in the Organisation of Economic Co-operation and Development (OECD) countries that essentially stymie most attempts at market development for food and cash crops in SSA and elsewhere in developing countries.

Effective research and extension services are essential to the implementation of improved N management. Innovations in extension and communication through NGOs, international and national research institutions, and the private sector are quite advanced in SSA and provide new approaches for working and communicating with farmers. Recognizing that socioeconomic constraints limit the short-term adoption of fertilizer, a variety of organic and integrated organic and fertilizer management strategies have been developed. These technologies are spreading rapidly throughout Africa, but their adoption is limited by land and labor constraints and the fact that they require programs and trained personnel to promote them. It is also acknowledged that these organically based technologies alone will not solve the food insecurity issues in SSA and that the application of fertilizers is essential. Priority should be placed on providing sufficient inputs and working markets to revitalize the agricultural sector in SSA (Vanlauwe et al., Chapter 8, this volume).

Concerns for the environment are secondary when food security and poverty are the major issues. In such situations, there is a possibility that international agreements such as the United Nations (UN) Convention to Combat Desertification and the Kyoto Protocol of the UN Climate Change Convention could provide mechanisms to pay farming communities for environmental services, such as carbon sequestration and biodiversity conservation, which they provide to the global society. Such payments could be used to provide inputs and other incentives for improved agricultural systems.

Excess Nitrogen

Most OECD countries continue to support agriculture financially, leading to highly productive agriculture with intensive N fertilizer use and a multitude of environmental problems. Many of these countries are pursuing ways of reducing the negative impacts by implementing environmental polices that impact on agricultural practices. The political will to reduce emissions varies, as have the impacts of their policies.

Politicians are faced with the problem that the environmental impacts from agriculture are displaced from the source of the pollution and have impacts at different geographic scales, depending on the type of emission. This has led to the formulation of policies at different scales (Table 5.5). Regulation and stimulation/incentives are two policy instruments for dealing with environmental problems. Regulation works by forbidding undesired practices and by penalizing transgressors; stimulation involves economic measures (subsidies, premiums, contracts, levies, taxes, tradable permits) and extension to promote desirable practices and discourage undesirable ones. Both types of policy instruments are typically used.

The costs of pollution are not borne by the farmer or the fertilizer industry. Society, therefore, has to impose incentives to reduce pollution. Legally, it is difficult to target the pollution directly, such as through a tax on the pollution, because this requires verification of the amount and type of pollution at farm or field level, the costs of which are prohibitive. Instead, policy needs to target indirect indicators of the pollution, such as N surplus or the amount of manure and fertilizer applied to individual crops. This can also be difficult to verify on an extensive basis.

Policies that involve economic measures are generally seen to lead to the most cost-effective reductions in pollution; this depends, however, on the type of economic incentive. Subsidizing organic farming, low N application, cover crops, and buffer strips can involve high costs relative to the pollution reductions, although other desirable effects on the landscape, biodiversity, and food safety may occur. Levies or taxes on N fertilizer or surplus are more flexible mechanisms but need to be designed to reduce undesirable effects. Levies and taxes on N surplus measured at the farm gate most directly target the pollution and give incentives for improving NUE. The financial incentives cannot ensure low N emissions unless all farmers in a country are severely restricted by these. In N-sensitive areas, financial measures are therefore often combined or substituted with regulation (e.g., maximum N rates, restrictions on crop choice, compulsory use of cover crops). Such regulations can be effective in reducing N pollution, but not all lead to increased NUE because many are directed toward land-use changes rather than changes in fertilizer use.

Environmental policies for N most often involve pollution-reduction targets and often result in policies that increase NUE because this will be reflected in a reduced N input. Such policies may work either directly through support of new and efficient technologies or indirectly through restricting the amount of N used or the N surplus

Table 5.5. Examples of policies to reduce negative environmental impacts of different nitrogen emissions and their scale of impact and political commitment

N loss type	Impact	Policy examples
Nitrates	Groundwater quality (regional)	EU, Nitrates Directive (nitrate sensitive areas) The Netherlands (national)
	Surface waters (regional)	EU, Water Framework Directive (catchment) USA, Clean Water Act (state) Denmark, Aquatic Action Plans (national)
Ammonia	Eutrophication (trans-national)	Europe, Gothenburg Protocol (national) Europe, Habitat Directive (local/regional)
Nitrous oxide	Global warming (global)	UNFCCC, Kyoto Protocol (national)

EU, European Union; UNFCC, United Nations Framework Convention on Climate Change.

obtained. New technologies may need to be subsidized during introduction; these subsidies may be removed later because widespread implementation reduces the costs of such technologies.

Policies for effectively reducing N losses involve large transaction costs for farm-level N accounting and public auditing of farm N accounts, farm inspection, and administration of financial incentives. Because most countries that have environmental problems associated with agriculture also subsidize agriculture, it might be argued that the individual farms should bear the farm-level costs and the government should assume the public costs.

Conclusions

The N fertilizer accessibility and use situations have arisen primarily from a suite of national and international policies in agriculture and environmental sectors. Currently, closer interaction among these sectors is occurring in Europe and to some extent in the United States in striving to balance food production and pollution concerns. When we look to the future, can countries that are beginning to intensify their agricultural production learn from the experiences of others? Are there measures that can be taken to stave off serious environmental problems before they occur? Where environmental problems do exist, how can future policies be crafted to solve these problems while ensuring food security and boosting economies?

Effective policies that focus on food security and rural development must ensure sufficient infrastructures that provide access to fertilizers and other inputs and to deliver products to the national and international markets. Provision of financial assistance and knowledge through education, research, and extension are also critical. Effective uptake of improved agricultural production will also probably initially require some type of subsidy for fertilizers or crop price supports. Developing countries should not refrain from supporting fertilizer use because of fear of environmental pollution. Currently, the risk is greater that the agricultural production potential of the land will be decreased through soil degradation resulting from insufficient fertilizer use. Nevertheless, there needs to be an awareness of the potential environmental effects so that regulations can be put in place at the appropriate time. Such environmental concerns should now be addressed in many parts of Asia but are probably a couple of decades away in much of SSA.

In areas where environmental problems due to excessive N use exist, future policies must be carefully constructed to ensure that past environmental gains are preserved and flexible enough to accommodate new knowledge. A portfolio approach ensures that a suite of options exists to address individual situations as well as multifaceted problems. Policies targeting N management practices must be in line with guidelines for free trade.

Agriculture must provide not only food and fiber for a growing world population but also a number of other ecosystem services, including clean air and water and preservation of biodiversity and landscapes. This concept requires finding the appropriate balance among the environmental, social, and economic functions in different regions. A multifunctional agriculture can be implemented only through closely integrated policies within agriculture, environment, research, and education. It will also require a close collaboration between public and private sector participants.

Literature Cited

Abdulrachman, S., Z. Susanti Pahim, A. Djatiharti, A. Dobermann, and C. Witt. 2004. Site-specific nutrient management in intensive irrigated rice systems of West Java, Indonesia. Pp. 171-192 in *Increasing productivity of intensive rice systems through site-specific nutrient management,* edited by A. Dobermann, C. Witt, and D. Dawe. Enfield, N.H.: Science Publishers, Inc., and Los Baños, Philippines: International Rice Research Institute.

Claassen, R. 2003. Emphasis shifts in U.S. agri-environmental policy. http://www.ers .usda.gov/AmberWaves/.

Claassen, R., L. Hansen, M. Peters., V. Breneman, M. Weinberg, A. Cattaneo, P. Feather, D. Gadsby, D. Hellerstein, J. Hopkins, P. Johnston, M. Morehart, and M. Smith.

2001. Agri-environmental policy at the crossroads: Guideposts on a changing landscape. Agric. Econ. Report No. 794 (Jan. 2001). United States Department of Agriculture Economics Research Service. http://www.ers.usda.gov/publications/aer794/.

De Clercq, P., A. C. Gertsis, G. Hofman, S. C. Jarvis, J. J. Neeteson, and F. Sinabell. 2001. *Nutrient management legislation in European countries*. Wageningen, The Netherlands: Wageningen Press.

FAO. 1999. *A fertilizer strategy for Zimbabwe*. Rome, Italy: Food and Agriculture Organization of the United Nations.

FAOSTAT. (United Nations Food and Agricultural Organization). 2004. *FAO agricultural data bases are obtainable on the world wide web*: http://www.fao.org.

Howarth, R. W., E. W. Boyer, W. J. Pabich, and J. N. Galloway. 2002. Nitrogen use in the United States from 1961–2002 and potential future trends. *Ambio* 31:88–96.

Humphreys, A. J. 1991. *The Zimbabwean fertiliser market*. Proceedings of an IFA regional conference for Sub-Saharan Africa. Harare, Zimbabwe: International Fertiliser Association.

Kichel, A. N., and C. H. B. Miranda. 1999. *Recuperação e renovação de pastagens degradadas. Recomendações Técnicas para Produção de Gado de Corte*. Technical Bulletin. Gado de Corte, Campo Grande, Brazil: Embrapa.

Kroeze, C., R. Aerts, N. van Breemen, D. van Dam, K. van der Hoek, P. Hoefschreuder, M. Hoosbeek, J. de Klein, H. Kros, H. van Oene, O. Oenema, A. Tietema, R. van der Veeren, and W. de Vries. 2003. Uncertainties in the fate of nitrogen. I: An overview of sources of uncertainty illustrated with a Dutch case study. *Nutrient Cycling in Agroecosystems* 66:43–69.

Kyllingsbæk, A. 2000. *Kvælstofbalancer og kvælstofoverskud i dansk landbrug 1979–1999*. DJF Rapport Markbrug no. 36, Danish Institute of Agricultural Sciences, Foulum, Denmark.

Mahmood, T., K. A. Malik, S. R. S. Shamsi, and M. I. Sajjad. 1998. Denitrification and total N losses from an irrigated sandy clay loam under maize-wheat cropping system. *Plant and Soil* 199:239–250.

Minnesota Department of Agriculture. 2003. *New program removes financial risk of implementing nutrient best management practices*. News Release 21 Jan. 2003. http://www.mda.state.mn.us/newsroom.htm.

Oenema, O. 2004. Governmental policies and measures regulating nitrogen and phosphorus from animal manure in European agriculture. *Journal of Agricultural Sciences* (in press).

Oliveira, O. C., I. P. de Oliveira, E. Ferreira, B. J. R. Alves, C. H. B. Miranda, L. Vilela, S. Urquiaga, and R. M. Boddey. 2001. Response of degraded pastures in the Brazilian Cerrado to chemical fertilization. *Pasturas Tropicales* 23:14–18.

Sanchez, P. A. 2002. Soil fertility and hunger in Africa. *Science* 295:2019–2020.

Smaling, E. M. A., S. M. Nandwa, and B. H. Janssen. 1997. Soil fertility is at stake. Pp. 47–61 in *Replenishing soil fertility in Africa*, edited by R. J. Buresh, P. A. Sanchez, and F. Calhoun. Madison, Wisconsin: Soil Science Society of America.

Tripathi, B. P., J. K. Ladha, J. Timsina, and S. R. Pascua. 1997. Nitrogen dynamics and balance in intensified rain-fed lowland rice-based cropping systems. *Soil Science Society of America* 61:812–821.

USDA and USEPA (United States Department of Agriculture and United States Environmental Protection Agency). 1999. *Unified national strategy for animal feeding operations* (March 9, 1999). http://www.epa.gov/npdes/pubs/finafost.pdf.

Zhu, Z. L., and Q. X. Wen. 1992. *Nitrogen in soils of China.* Nanjing: Jiangsu Science and Technology Publishing House.

Zimmer, H., and K. Euclides-Filho. 1997. Brazilian pasture and beef production. Pp. 1–9 in *Proceedings International Symposium on Animal Production Under Grazing 4–6 Nov.*, edited by J. A. Gomide. Viçosa, Brazil: Universidade Federal de Viçosa, Brazil.

PART III
Low-input Systems

6

Improving Fertilizer Nitrogen Use Efficiency Through an Ecosystem-based Approach

Laurie E. Drinkwater

The application of ecologic concepts to agriculture has become increasingly widespread since the mid-1970s. In some areas of agricultural management, such as pest and weed control, the application of ecologic principles in research have become commonplace, resulting in management options that blend biologically and chemically based strategies (Lewis et al. 1997; Liebman and Gallandt 1997). These approaches have drawn heavily on population and community ecology and have resulted in reduced chemical use through successful application of biological controls (Lewis et al. 1997). They have also been instrumental in the success of organic production systems where the goal is to rely primarily on biological processes with only minimal use of nonsynthetic pesticides. An integrated nutrient management strategy based on ecologic concepts has yet to be broadly applied; however, this approach could make significant contributions to improving the efficiency of nitrogen use (NUE).

Nutrient Management as Applied Ecology

Nutrient management falls within the purview of ecosystem ecology, which aims to understand biogeochemical processes at multiple scales. Application of an ecosystem framework to agriculture expands the scope of the current agronomic framework to include management of biogeochemical processes in addition to crop assimilation of N (Table 6.1). Rather than focusing solely on soluble, inorganic plant-available pools, an ecosystem-based approach seeks to optimize organic and mineral reservoirs with longer mean residence times (MRTs) that can be accessed through microbial- and plant-mediated processes. This requires deliberate use of varied nutrient sources and strategic increases in plant diversity to restore desired agroecosystem functions such as nutrient and soil retention, internal cycling capacity, or aggregation. Breeding for cultivars and their associated microorganisms that do not require surplus nutrient additions is critical if plant and microbial-mediated ecosystem processes such as mineralization–immobilization, biological weathering, and carbon sequestration are to be harnessed. Integrated management of biogeochemical processes that regulate the cycling of nutrients and carbon (C), combined with increased reservoirs that are more readily retained in the soil, will greatly reduce the need for surplus nutrient additions.

Nitrogen Use Efficiency in Organically Managed Systems

Organic agriculture is an example of one approach that draws heavily from this framework. Organic farmers seek to manage plants, soil organic matter (SOM), and soil organisms to maintain internal cycling capacity (Howard 1945). This view is the basis for identifying the soil fertility management practices used in organic agriculture. Management practices are geared toward maintaining soil N stored as SOM rather than supplying plant-available fertilizers directly to crops each growing season (cf. Organic Farming Research Foundation 2002). The intention is to manage C in conjunction with the full range of soil organic N reservoirs, particularly those with relatively long MRTs that can be accessed by crops via microorganisms. Three soil fertility management practices are typically used in organic cropping systems that determine the cycling and availability of N in the soil: (1) the use of organic residues and biological N fixation as N sources; (2) living plant cover maintained as much as possible with cover crops, relay cropping, and intercropping; (3) plant species that are diversified in space and time to fulfill a variety of functions. Ideally, N inputs from N-fixing crops balance N removed as harvested exports. Consistent use of all three is unique to organic cropping systems, as is the prohibition of synthetic inorganic N. Whereas many conventional systems may use one or more of these practices, they are not generally considered important tools for improving NUE. Application of an ecosystem-based framework does not preclude the use of inorganic N fertilizers; indeed, in some instances, use of these sources may be more ecologically and economically sound than use of organic N forms.

Our understanding of the impact of organic management on agroecosystem-scale NUE is based on a few long-term systems experiments. These comparative system experiments suggest that organic systems have the potential to achieve a more favorable balance of inputs and exports compared with conventionally managed systems while achieving comparable yields for most crops (Clark et al. 1998; Drinkwater et al. 1998; and Mäder et al. 2002). One study comparing organic and conventional grain production found that after 15 years a larger proportion of total N inputs could be accounted for, either as harvested exports or in SOM pools in the organically managed rotations (Drinkwater et al. 1998). In this experiment, soil N and corn yields in an organic grain rotation where N inputs and exports were close to balanced did not differ significantly from the conventional system. In cases where surplus N was added, a significant proportion of that surplus could be accounted for by increases in N stored

Table 6.1. Characteristics of the current nutrient agronomic framework (Balasubramanian et al., Chapter 2, this volume; Havlin, Chapter 12, this volume; Dobermann and Cassman, Chapter 19, this volume) compared with an ecosystem-based approach (agroecosystem framework)

	Agronomic framework	Agroecosystem framework
Goals	Maximize crop uptake of applied N to achieve yield goal and reduce environmental losses	Achieve optimal yields and maintain soil reservoirs while balancing N additions and exports as much as possible
Nutrient management strategy	Manage crop to create a strong sink for fertilizer-N by removing all growth limiting factors & by providing an optimum delivery system	Manage agroecosystem to increase internal N cycling capacity to (1) maintain N pools that can be accessed through plant- and microbially mediated processes and (2) conserve N by creating multiple sinks in time and space for inorganic N
Nitrogen pools actively managed	Inorganic N	All N pools, organic and inorganic
Processes targeted by nutrient management	Crop uptake of N	Plant and microbial assimilation of N, C cycling, N and C storage, other desirable N transformations
Strategy toward microbially mediated N transformations	Manage to eliminate or inhibit as much as possible	Manage to promote N transformations that conserve N, reduce transformations that lead to losses by maintaining small inorganic N pools
Strategies for reducing NO$_3$ leaching	Increase crop uptake of added N, use chemicals that inhibit nitrification	Minimize inorganic pool sizes through management of multiple processes
Assessment of NUE	Metrics reflect fertilizer uptake of the crop, time step of metrics is one growing season	Metrics based in budgeting framework, reflect N balance and yield, time-step flexible
Typical experimental approaches	Short-term, small-plot, empirical, factorial experiments dominate	Hypothesis-driven systems and factorial experiments within an agroecosystem context, spatial and temporal scales set by processes to be studied

NUE, nitrogen use efficiency.

as SOM (Clark et al. 1998; Drinkwater et al. 1998). The conventional systems receiving only inorganic N did not show a net accrual of soil N despite having a significant N surplus. Understanding the underlying mechanisms that enable some organically managed cropping systems to achieve high yields while reducing N losses will contribute to improving management of fertilizer N.

Agroecosystem Framework to Improve Nitrogen Use Efficiency

Clearly, the application of this strategy of managing biogeochemical processes to improve fertilizer use efficiency requires a somewhat different approach than one based solely on organic N sources. Recoupling N and C cycles is central to increasing internal N cycling capacity, which is one important component of this agroecosystem framework (Table 6.1). Increasing the capacity of the soil to supply N can improve NUE through reductions in the amount of fertilizer N that must be applied for each crop. As requirements for inorganic N are reduced, NUE generally increases (Dobermann and Cassman, Chapter 19, this volume). Studies of integrated cropping systems where organic and inorganic fertilizers are used simultaneously show that diversifying N sources improves NUE (Olesen et al., Chapter 9, this volume; Vanlauwe, Chapter 8, this volume). Reliance on diverse organic N sources through the use of recycled organic residues and return of crop residues serves as an important means of maintaining the various pools of SOM in agricultural systems; however, other practices can also contribute to building SOM pools. Furthermore, it is also important to consider how increased SOM pools impact plant- and microbial-mediated processes regulating C and N cycling. The remaining discussion focuses on three key areas that appear to have the largest potential for improving field-scale fertilizer NUE while maintaining yields.

Cover Crops and Rotational Diversity

Simplified rotations became possible when the use of synthetic fertilizers and chemical weed controls eliminated the need for cover crops and forages (Auclair 1976). These crops have little or no cash value per se but help to maintain internal cycling capacity through a variety of mechanisms. For example, the preferential removal of winter annuals from large expanses of agricultural lands has increased the prevalence of bare fallows. This reduction in the time frame of living-plant cover reduces C fixation and increases erosion and depletion of SOM stocks (Aref and Wander 1998; Campbell and Zentner 1993). Reduced levels of SOM combined with the absence of plant activity during extended periods of time increases the susceptibility of these ecosystems to N saturation and resulting N losses (Fenn et al. 1998).

The use of cover crops and relay crops in annual rotations improves temporal synchronization of N mineralization and N uptake, leading to significant reductions in N leaching (McCracken et al. 1994). Replacement of bare fallows with cover crops increases SOM pools, creating a positive feedback that permits fertilizer N additions to the cash crop to be reduced as the capacity of the soil to provide N through mineralization increases. Extending the time frame of plant growth and rhizodeposition supports the soil microbial community for a greater part of the year and may also increase the proportion of C retained by the microbial biomass (Jans-Hammermeister et al. 1998). Thus, the use of cover crops increases SOM levels and improves NUE through multiple mechanisms.

Most research on cover crops has been conducted in short-term experiments that have focused on yield and to a lesser extent on leaching measurements. Studies of these feedbacks resulting from increased recycling of fertilizer N and increased levels of labile SOM have not been conducted, so the impact of cover crops on fertilizer requirements under steady-state conditions is not known.

Other changes in C and N cycling are related to specific differences in plant species characteristics, such as biochemical composition of litter and root exudates. For example, plant species often used as cover crops, such as small grains and legumes, tend to promote soil aggregation to a larger extent than crops like corn (Haynes and Beare 1997; Tisdall and Oades 1979). Clear differences in the fate of root-derived C from different plant species seem to be partially responsible for differences in aggregation formation. The inclusion of plant species that foster aggregate formation may contribute to the differential retention of C observed in some long-term studies (Drinkwater et al. 1998; Puget and Drinkwater 2001) and may reduce decomposition and net N mineralization in the absence of plant roots.

Plant–Microbial Interactions

Plant–microbial interactions regulate a wide range of biogeochemical processes (Haynes and Beare 1997; Hooper and Vitousek 1997). Management of the basic exchange of C from primary producers for nutrients from decomposers has not been attempted in agroecosystems, despite the opportunity afforded by the rhizosphere as the site of this mutual codependency between decomposers and plants (Wall and Moore 1999). Plants can stimulate decomposition of organic substrates by supplying labile C to decomposers in the rhizosphere (Clarholm 1985). The rate of decomposition and N mineralization varies with plant species (Cheng et al. 2003), rhizosphere community composition (Chen and Ferris 1999; Clarholm 1985; Ferris et al. 1998) and nutrient availability (Liljeroth et al. 1994; Tate et al. 1991). This exchange does not depend simply on net N mineralization during decomposition (Clarholm 1985). Instead, the release of nutrients for plant uptake appears to be dependent on the involvement of secondary consumers feeding on the primary decomposers (Clarholm 1985) because of differences in the stoichiometry between the two trophic levels (Chen and Ferris 1999; Ferris et al. 1998). There is growing evidence that plants can influence the rate of net N mineralization based on their need for nutrients by modifying the amount of soluble C excreted into the rhizosphere (Hamilton and Frank 2001).

Increased reliance on this exchange of C for N in organic systems may explain why yields comparable to conventional systems can be achieved in organic systems where inorganic N pools are only 2 to 3 mg kg^{-1} soil. The tight coupling between net mineralization of N and plant uptake in the rhizosphere reduces the potential for N losses. Inorganic nutrient pools can be extremely small, whereas high rates of plant production are maintained if N mineralization and plant assimilation are spatially and temporally connected in this manner (Jackson et al. 1988). To manage this process effectively, many questions remain to be answered. In particular, it is important to know (1) which SOM pools are being accessed by plant-mediated decomposition and (2) how to manage N fertilizers to increase these pools while minimizing SOM pools that can be mineralized in the absence of plants and that would contribute to potential losses of N. Other aspects, such as food-web structure, could be influenced by management to optimize this process.

Microbial-mediated Processes

Microorganisms represent a substantial portion of the standing biomass in agricultural ecosystems and contribute to the regulation of C sequestration, N availability, and losses and P dynamics. For cultivated systems, the N in soil prokaryotes is estimated to be an average of 630 kg ha^{-1} in the first meter of soil (Whitman et al. 1998). This is a significant N pool, and increased understanding of the decomposer community could be used to develop management strategies that enhance the flux of C and N through this pool to improve NUE. It is clear that the size and physiologic state of the standing microbial biomass is influenced by management practices, including rotational diversity (Anderson and Domsch 1990), tillage (Holland and Coleman 1987), and the quality and quantity of C inputs to the soil (Fließbach and Mäder 2000; Lundquist et al. 1999).

Microbial community composition and metabolic status determine the balance between C released through respiration and C assimilation into biomass during decomposition as well as the biochemical composition of that biomass. Changes in microbial community structure can lead to increased C retention if the management practices result in fungal-dominated decomposer communities (Holland and Coleman 1987). Decomposers in soils with greater diversity of plant species (Anderson and Domsch 1990) or larger abundance of C relative to N (Aoyama et al 2000; Fließbach et al. 2000) have reduced energy requirements for maintenance and therefore convert a greater proportion of metabolized C to biomass. These studies, however, did not characterize microbial community composition (Fließbach et al. 2000).

Finally, it would also be advantageous to manage bacterial metabolic pathways for several key transformations of N. Denitrification contributes to significant N losses from agricultural systems and is regulated by a complex array of environmental factors. Recent studies of denitrifier populations from different soils suggest that management-induced changes in the soil environment alter both the composition and functional characteristics of the denitrifier community (Cavigelli and Robertson 2000, 2001). Denitrifiers from the intensively managed soil were more sensitive to O_2 levels and produced a greater proportion of N_2O compared with denitrifiers from an early successional plant community. This evidence supporting the connection between management and the characteristics of a microbial functional group suggests that it may be possible to reduce N losses through manipulation of microbial functional groups that control N transformations.

A second anaerobic pathway, dissimilatory nitrate reduction to ammonium (DNRA), was recently found to occur in a broad range of unmanaged terrestrial ecosystems (Silver et al. 2001). Previously this process was thought to be limited to extremely anaerobic, C-rich environments, such as sewage sludge and submerged sediments, including flooded rice soils (Maier et al. 1999). Silver et al. (2001) reported that average rates of DNRA were threefold greater than denitrification in upland tropical forest soils and concluded that the resulting reduction in NO_3 availability to denitrifiers and leaching may contribute to N conservation in these ecosystems. The presence of microbes capable of DNRA in upland agricultural systems has yet to be determined, but there is no reason to expect this process to be excluded from these ecosystems. Organically managed soils have larger pools of labile C (Drinkwater et al. 1995; Wander et al. 1994) and also have a wider NH_4:NO_3 ratio compared with soils managed with inorganic fertilizers (Drinkwater et al. 1995), thus suggesting that in agroecosystems with increased C availability, DNRA could be an important N-conserving process that could be enhanced through management.

Conclusions

Agricultural research has made significant contributions toward understanding the mechanisms regulating soil biogeochemical processes such as N and C cycling. Application of an ecosystem-based conceptual model to nutrient management offers the opportunity to apply this understanding fully to improving NUE. This approach provides a unifying framework that is particularly well suited to characterizing interrelationships among the environmental conditions (abiotic components), management practices, and biogeochemical processes that control yield, NUE, carbon storage, and N losses. The use of a unifying conceptual model will also improve the efficiency of research by ensuring that a cohesive body of knowledge is generated, regardless of the spatial and temporal scales that define the boundaries of individual nutrient management studies (Drinkwater 2002). This is particularly critical at this juncture because further improvement of NUE will require a greater understanding of the ecosystem processes governing the fate of all forms of added N within agroecosystems (Havlin,

Chapter 12, this volume; Dobermann and Cassman, Chapter 19, this volume). Finally, intentional management of ecosystem processes will reconcile production and environmental goals by promoting the development of NUE indices that reflect the capacity of cropping systems to retain N while optimizing yields.

Literature Cited

Anderson, J. P. E., and K. H. Domsch. 1990. Application of eco-physiological quotients (qCO2 and qD) on microbial biomasses from soils of different cropping histories. *Soil Biology and Biochemistry* 22:251–255.

Aoyama, M., D. A. Angers, A. N'Dayegamiye, and N. Bissonnette. 2000. Metabolism of ^{13}C-labeled glucose in aggregates from soils with manure application. *Soil Biology and Biochemistry* 32:295–300.

Aref, S., and M. M. Wander. 1998. Long-term trends of corn yield and soil organic matter in different crop sequences and soil fertility treatments on the Morrow Plots. *Advances in Agronomy* 62:153–197.

Auclair, A. N. 1976. Ecological factors in the development of intensive management ecosystems in the midwestern USA. *Ecology* 57:431–444.

Campbell, C. A., and R. P. Zentner. 1993. Soil organic matter as influenced by crop rotations and fertilization. *Soil Science Society of America Journal* 57:1034–1040.

Cavigelli, M. A., and G. P. Robertson. 2000. The functional significance of denitrifier community composition in a terrestrial ecosystem. *Ecology* 81:1402–1414.

Cavigelli, M. A., and G. P. Robertson. 2001. Role of denitrifier diversity in rates of nitrous oxide consumption in a terrestrial ecosystem. *Soil Biology and Biochemistry* 33:297–310.

Chen, J., and H. Ferris. 1999. The effects of nematode grazing on nitrogen mineralization during fungal decomposition of organic matter. *Soil Biology and Biochemistry* 31:1265–1279.

Cheng, W., D. W. Johnson, and S. Fu. 2003. Rhizosphere effects on decomposition: Controls of plant species, phenology, and fertilization. *Soil Science Society of America Journal* 67:1418–1427.

Clarholm, M. 1985. Interactions of bacteria, protozoa and plants leading to mineralization of soil nitrogen. *Soil Biology and Biochemistry* 17:181–187.

Clark, M. S., W. R. Horwath, C. Shennan, and K. M. Scow. 1998. Changes in soil chemical properties resulting from organic and low-input farming practices. *Agronomy Journal* 90:662–671.

Drinkwater, L. E. 2002. Cropping systems research: Reconsidering agricultural experimental approaches. *HortTechnology* 12:355–361.

Drinkwater, L. E., D. K. Letourneau, F. Workneh, B. A. H. C. Van, and C. Shennan. 1995. Fundamental differences between conventional and organic tomato agroecosystems in California. *Ecological Applications* 5:1098–1112.

Drinkwater, L. E., P. Wagoner, and M. Sarrantonio. 1998. Legume-based cropping systems have reduced carbon and nitrogen losses. *Nature* 396:262–265.

Fenn, M. E., M. A. Poth, J. D. Aber, J. S. Baron, B. T. Bormann, D. W. Johnson, A. D. Lemly, S. G. McNulty, D. F. Ryan, and R. Stottlemyer. 1998. Nitrogen excess in North

American ecosystems: Predisposing factors, ecosystem responses, and management strategies. *Ecological Applications* 8:706–733.

Ferris, H., R. C. Venette, H. R. van der Meulen, and S. S. Lau. 1998. Nitrogen mineralization by bacteria-feeding nematodes: Verification and measurement. *Plant and Soil* 203:159–171.

Fließbach, A., and P. Mäder. 2000. Microbial biomass and size-density fractions differ between soils of organic and conventional agricultural systems. *Soil Biology and Biochemistry* 32:757–768.

Fließbach, A., P. Mäder, and U. Niggli. 2000. Mineralization and microbial assimilation of ^{14}C labeled straw in soils of organic and conventional agricultural systems. *Soil Biology and Biochemistry* 32:1131–1139.

Hamilton, E. W. III, and D. A. Frank. 2001. Can plants stimulate soil microbes and their own nutrient supply? Evidence from a grazing-tolerant grass. *Ecology* 82:2397–2402.

Haynes, R. J., and M. H. Beare. 1997. Influence of six crop species on aggregate stability and some labile organic matter fractions. *Soil Biology and Biochemistry* 29:1647–1653.

Holland, E. A., and D. C. Coleman. 1987. Litter placement effects on microbial and organic matter dynamics in an agroecosystem. *Ecology* 62:425–433.

Hooper, D. U., and P. M. Vitousek. 1997. The effects of plant composition and diversity on ecosystem processes. *Science* 277:1302–1305.

Howard, A. 1945. *Farming and gardening for health or disease.* London: Faber and Faber.

Jackson, L. E., J. P. Schimel, and M. K. Firestone. 1989. Short-term partitioning of ammonium and nitrate between plants and microbes in an annual grassland. *Soil Biology and Biochemistry* 21:409–415.

Jans-Hammermeister, D. C., W. B. McGill, and R. C. Izaurralde. 1998. Management of soil C by manipulation of microbial metabolism: Daily vs. pulsed C additions. *Advances in Soil Science* 11:321–333.

Lewis, W. J., L. J. C. Van, S. C. Phatak, and J. H. Tumlinson, III. 1997. A total system approach to sustainable pest management. *Proceedings National Academy of Sciences* 94:12243–12248.

Liebman, M., and E. R. Gallandt. 1997. Many little hammers: Ecological management of crop-weed interactions. Pp. 291–343 in *Ecology in agriculture*, edited by L. E. Jackson. New York: Academic Press.

Liljeroth, E., P. Kuikman, and J. A. Van Veen. 1994. Carbon translocation to the rhizosphere of maize and wheat and influence on the turnover of native soil organic matter at different soil nitrogen levels. *Plant and Soil* 161:233–240.

Lundquist, E. J., L. E. Jackson, K. M. Scow, and C. Hsu. 1999. Changes in microbial biomass and community composition, and soil carbon and nitrogen pools after incorporation of rye into three California agricultural soils. *Soil Biology and Biochemistry* 31:221–236.

Mäder, P., A. Fließbach, D. Dubois, L. Gunst, P. Fried, and U. Niggli. 2002. Soil fertility and biodiversity in organic farming. *Science* 296:1694–1697.

Maier, R. M., I. L. Pepper, and C. P. Gerba. 1999. *Environmental microbiology.* New York: Academic Press.

McCracken, D. V., M. S. Smith, J. H. Grove, C. T. MacKown, and R. L. Blevins. 1994. Nitrate leaching as influenced by cover cropping and nitrogen source. *Soil Science Society of America Journal* 58:1476–1483.

Organic Farming Research Foundation. 2002. http://www.ofrf.org/general/about_
organic/index.html

Puget, P., and L. E. Drinkwater. 2001. Short-term dynamics of root and shoot-derived
carbon from a leguminous green manure. *Soil Science Society of America Journal*
65:771–779.

Silver, W. L., D. J. Herman, and M. K. Firestone. 2001. Dissimilatory nitrate reduction
to ammonium in upland tropical forest soils. *Ecology* 82:2410–2416.

Tate III, R. L., R. W. Parmelee, J. G. Ehrenfeld, and L. O'Reilly. 1991. Nitrogen mineralization: Root and microbial interactions in pitch pine microcosms. *Soil Science Society of
America Journal* 55:1004–1008.

Tisdall, J. M., and J. M. Oades. 1979. Stabilization of soil aggregates by the root system
of ryegrass. *Australian Journal of Soil Research* 17:429–441.

Wall, D. H., and J. C. Moore. 1999. Interactions underground: Soil biodiversity, mutual-
ism, and ecosystem processes. *Bioscience* 49:109–117.

Wander, M. M., S. J. Traina, B. R. Stinner, and S. E. Peters. 1994. Organic and conventional management effects on biologically active soil organic matter pools. *Soil Science
Society of America Journal* 58:1130–1139.

Whitman, W. B., D. C. Coleman, and W. J. Wiebe. 1998. Prokaryotes: The unseen
majority. *Proceedings National Academy of Sciences* 95:6578–6583.

7

Nitrogen Dynamics in Legume-based Pasture Systems

M. B. Peoples, J. F. Angus, A. D. Swan, B. S. Dear, H. Hauggaard-Nielsen, E. S. Jensen, M. H. Ryan, and J. M.Virgona

Legumes are important components of temperate pastures used to produce wool, dairy, and meat in Australia, New Zealand, Western Europe, and both North and South America. In the case of Australia, inputs of symbiotically fixed N via volunteer or sown legume pastures also provide a major source of N for grain crops grown in phased rotations. Tropical pastures containing legumes are also sown in Brazil and in other parts of South America and Australia. The presence of 20 to 40 percent of forage legumes in tropical pastures increases meat and milk production by as much as 10-fold (Thomas 2000). Apart from the tropics of Australia, where *Stylosanthes* spp. and more recently butterfly pea (*Clitoria ternatea*) have been sown, and Brazil, where legume/grass pastures are now part of zero tillage systems, legumes have not been widely adopted. Data collected from a number of South American countries suggest that less than 10 percent of improved pastures included legumes (Thomas 2000); although forage legumes have been adopted by some commercial farmers in areas of Africa, however, the impact in smallholder agriculture has been small (Giller 2001).

Inputs of Fixed Nitrogen

A wide diversity of legume species are grown in temperate (Peoples and Baldock 2001) and tropical pastures (Giller 2001); however, comparative estimates of N_2 fixation across countries and regions are available for relatively few species. A summary of the range of measures of N_2 fixation obtained for commonly studied forage legumes is presented in Table 7.1.

Most modern methods used to quantify inputs of fixed N in pastures systems par-

Table 7.1. Estimates of the proportion (%Ndfa) and annual amounts of shoot nitrogen fixed by a selection of important temperate pasture legume species[1]

| | Shoot N fixed | | | |
| | %Ndfa | | $kg\ N\ ha^{-1}\ yr^{-1}$ | |
Species Country or region	Range measured	Common range	Range measured	Common range
Annual legumes				
Subterranean clover				
(*Trifolium subterraneum*)				
Australia	50–100	75–90	2–238	50–110
Annual medics				
(*Medicago* spp.)				
Australia	48–99	70–85	2–220	70–100
North America	72–86	80–83	101–205	125–140
Perennial legumes				
Lucerne / Alfalfa				
(*Medicago sativa*)				
Australia	25–93	65–75	4–284	80–140
North America	33–78	60–70	106–308	160–260
Europe	70–88	70–80	93–319	100–250
White clover				
(*Trifolium repens*)				
Australia	58–94	65–85	11–236	40–125
New Zealand	45–76	55–70	65–291	80–180
Europe	79–94	80–90	15–283	100–220

[1] Collated from data published by Ledgard and Steele (1992), Jørgensen et al. (1999), Vinther and Jensen (2000), and Peoples and Baldock (2001). Data from North America include Canada and the United States, while the European data include results from Austria, Denmark, Sweden, Switzerland, and the United Kingdom.

tition legume N into that proportion derived from atmospheric N_2 (percent of Ndfa) and that coming from the soil. The amounts of N_2 fixed over a period of growth are calculated as the product of percent of Ndfa and measures of legume N accumulation, usually determined from foliage dry matter (DM) and N content. Almost all published data on N_2 fixation have been based on such measures of aboveground biomass N because analyses of roots physically recovered from soil often suggest that roots contain only a small fraction (<15 percent) of the total plant N. Recent studies using a range of ^{15}N-based techniques, however, now indicate that N either associated with, or derived from, the turnover and decomposition of nodulated roots of pasture legumes can represent 40 to 70 percent of the total plant N (Peoples and Baldock 2001). Consequently, total inputs of fixed N might be twice the amount calculated from the more traditional shoot-based measurements presented in Table 7.1.

Whereas some pastures exist as pure legume swards, most contain mixtures of legumes, grasses, and broadleaf species. Competition between these species results in low concentrations of available soil N for much of the growing season. As a consequence, legume reliance on N_2 fixation for growth (percent of Ndfa) tends to be high (70–90 percent) for annual temperate legumes (Table 7.1) and most tropical species (Giller 2001). Perennials such as Lucerne (alfalfa) have a greater capacity to scavenge soil mineral N than annuals (Dear et al. 1999), and the percent Ndfa is often lower than that observed for annual legumes (Table 7.1).

The data presented in Table 7.1 and elsewhere (Giller 2001) suggest potential inputs of several hundred kilograms of fixed N per year for most temperate and tropical legume species, with reported peak rates of N_2 fixation greater than 3 kg N ha^{-1} d^{-1} (Jørgensen et al. 1999). Many temperate (Peoples and Baldock 2001) and tropical (Thomas et al. 1997) forage legumes fix between 20 and 25 kg shoot N with each tonne of legume foliage DM accumulated. White clover seems to be an exception to this generalization because it often fixes around 40 kg of shoot N per tonne DM as a result of the high N concentrations (4.5–6 percent) in its foliage (Vinther and Jensen 2000).

Variations in legume biomass and N_2 fixation can result from differences in pasture legume contents or composition (Dear et al. 1999), environmental limitations on legume productivity or seedling establishment (particularly water availability), or nutritional constraints to legume growth, especially soil acidity and aluminium toxicity (Peoples and Baldock 2001). Low phosphorus availability can be a particular problem in tropical soils, and nodulation and growth on phosphorus-deficient soils have been reported to be stimulated by mycorrhizal inoculation (Giller 2001). Fertilizer N tends to be applied only to intensive clover/grass dairy pastures, but where it is used, it can suppress N_2 fixation through its influence on nodulation, the percent of Ndfa, and clover growth (Whitehead 1995).

Differences in annual inputs of fixed N can also depend on whether the pasture is based on annual or perennial legumes. Because perennials can grow and fix N under conditions unsuitable for either the establishment or growth of annual legumes, there is more consistent production of legume biomass and N_2 fixation compared with the extreme year-to-year variations observed in most annual pastures (Peoples et al. 1998).

Soil Nitrogen Dynamics Under Pasture

Many reports of progressive improvements of soil N status under legume-based pastures have appeared. Annual increments of total soil organic N of 30 to 80 kg N ha^{-1} are common beneath both temperate and tropical legume-based annual pastures, but average rates of organic N accretion greater than 100 kg N ha^{-1} yr^{-1} (0–2m) can occur in soils under Lucerne (Peoples and Baldock 2001).

Because legume residues tend to have a low C:N ratio, they are usually expected to result in net mineralization. A range of other constituents also can influence microbial activity and mineralization, however, and predictions based simply on the C:N ratio of tissue can sometimes be misleading. For example, the incubation studies of Bolger et al. (2003) demonstrated that transient immobilization of N can be greater in soil containing Lucerne root residues than in soil amended with subterranean clover root residues, despite a near identical total C:N ratio. The immobilization induced by the Lucerne material appeared to be related to larger amounts of dry matter and C in the readily labile soluble fraction, which presumably stimulated initial microbial activity and demand for N (Bolger et al. 2003). The rate of release of N was also shown to be much higher from leaf litter of the tropical species *Macroptilium atropurpeum* than from *Desmodium intortum,* even though the N contents were similar (Thomas and Asakawa 1993). In this case the difference was attributed to the greater concentrations of polyphenols in the *Desmodium,* which complexed with leaf proteins to render them resistant to microbial attack.

Most studies investigating factors that influence mineralization processes tend to focus on the residues from a single species. Whereas legumes are sometimes grown in pure swards, more usually they will be growing in association with other species. The impact of residues of different chemical quality on mineralization was recently investigated in a series of incubation studies that compared the N dynamics of different mixtures of Balansa clover (*Trifolium michelianum*) and Italian ryegrass (*Lolium multiflorum*) shoot residues (Figure 7.1). Concentrations of total inorganic N at the end of the incubation period were highest in the 100 percent clover treatment and declined with decreasing amounts of clover and increasing additions of ryegrass (Figure 7.1). Net mineralization was measured in all treatments where leguminous material represented more than 25 percent of the total residues (Figure 7.1). Because the clover residues were labeled with ^{15}N, it was possible to determine that only a proportion of this additional mineral N was derived from the original leguminous materials (Figure 7.1). This implies that the presence of clover residues stimulated mineralization of native soil organic N. The observed concentrations of inorganic N were comparable to values predicted from simple arithmetic calculations based on the 100 percent clover and 100 percent ryegrass treatments (Figure 7.1). A similar study with white clover and ryegrass also demonstrated that net N mineralization was closely linked with the C:N ratio and N contents of the added material and reflected the patterns of N turnover of the individual residue components (de Neergaard et al. 2002).

Whereas the preceding discussion focused on the accumulation of soil mineral N via decomposition of organic residues, rhizodeposition of N released from living roots and the return of foliage N ingested by grazing animals in urine and dung are additional pathways of transfer for legume N in pasture systems. Research data collected from intensively grazed dairy pastures (Ledgard and Steele 1992) suggest that the annual amounts of legume N transferred via excreta to accompanying non-leguminous species

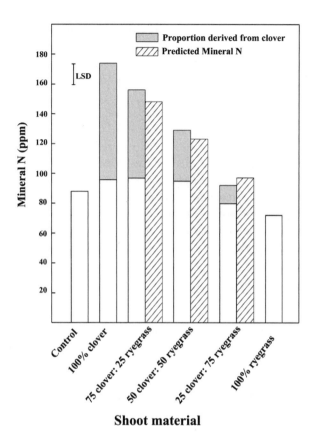

Shoot material

Figure 7.1. The effect of mixing shoot residues derived from balansa clover (C:N = 12.1) and Italian ryegrass (C:N = 48.4) on the accumulation of soil mineral N under controlled conditions (Hauggaard-Nielsen and Ryan, unpublished data). The nil-residue control provides a measure of N mineralization from the native soil organic matter. Data for the residue mixes were derived from 15 mg of dry shoot material added to the equivalent of 25 g dry weight of an acidic loam soil (Natric Palexeralf) compacted to a standard bulk density of 1.35 g/cm3, wetted to field capacity and incubated at 15°C for 141 days. The LSD bar indicates the least significant difference at the 5 percent level. The shaded section of each clover treatment indicates 15N-derived estimates of the proportion of the total mineral N that originated from the added legume residues. Concentrations of soil mineral N in the various mixtures predicted on the basis of the 100 percent clover and 100 percent ryegrass treatments are shown in the hatched histograms.

(60 kg N ha⁻¹) can be similar to the combined contribution from rhizodeposition and decomposition (70 kg N ha⁻¹, representing 26 percent N fixed and 27 percent of the grass N). Other estimates of N transfer in the absence of livestock commonly range from 1 to 20 percent of the N fixed, satisfying 20 to 40 percent of the grass N requirements (Jørgensen et al. 1999).

Soils under annual temperate pastures are characterized by yearly cyclic changes in soil mineral N (<20 to >250 kg N ha^{-1} in the top 1 to 2 m; Peoples and Baldock 2001). Levels are usually highest in autumn to early winter and lowest during the peak periods of pasture growth and N demand in spring. Factors affecting the concentrations of soil inorganic N in autumn include the intensity of grazing, rainfall to stimulate microbial activity, or the presence of summer weeds that will both assimilate inorganic N and slow subsequent mineralization by drying the soil profile (Peoples and Baldock 2001).

Although annual legumes grow within most perennial pastures, such seasonal flushes of mineral N occur only when the perennial component (legume or grass) is present at low densities. In the case of Lucerne pastures in southeastern Australia, concentrations of soil mineral N generally remain below 60 kg N ha^{-1} (0–2 m) throughout much of the pasture phase because most of the N mineralized will tend to be assimilated by Lucerne roots. Consequently, autumn concentrations of soil nitrate tend to be lower beneath Lucerne pastures than under annual clover swards (Dear et al.1999).

Impact of Grazing Animals

Typically 75 to 95 percent of the foliage N ingested by livestock is returned to pastures as urine and feces. The proportion of land affected by excreta depends on the stocking rate, but commonly 4 to 10 percent of the pasture's surface might be covered by urine patches and feces following a single grazing by cattle or a flock of sheep (Jarvis et al. 1995). Annual returns of excreta can represent from 130 to 240 kg N ha^{-1} yr^{-1} for steers or sheep grazing clover/grass pastures to 300 to 450 kg N ha^{-1} yr^{-1} for a dairy herd on heavily N-fertilized grasslands (Jarvis et al. 1995). Depending on the N content of the feed, between 40 and 83 percent of this excreted N will be in the form of urine (Fillery 2001). Urine is returned to localized areas of pasture at rates equivalent to 100 to 400 kg N ha^{-1} by sheep and 600 to 1200 kg N ha^{-1} by cattle, whereas cattle dung pats may represent up to the equivalent of 2000 kg N ha^{-1} (Ledgard and Steele 1992). Such high rates of N supply to soil not only suppress N$_2$ fixation by the legume (Vinther 1998) but are also likely to exceed local plant demands and result in volatile or leaching losses of N.

The extent of gaseous losses (as a percent of N excreted) from urine tends to be four to five times greater than that from dung (Table 7.2) because of the higher concentrations of ammonium and nitrate under urine patches. Although losses from individual urine or dung patches can be large, the available data suggest that gaseous losses from grazed pastures may be modest simply because most of the land area is not directly affected by excreta (Table 7.2). Losses of urine N by ammonia volatilization can be expected to be greatest during periods of infrequent rainfall and high soil temperatures in both tropical and temperate regions (Fillery 2001). Nitrogen losses as the greenhouse gas N$_2$O during either nitrification or denitrification appear to be relatively small from grazed legume pastures (Table 7.2) and may represent only 5 to 50 percent of the emis-

Table 7.2. Summary of estimates of gaseous losses of nitrogen from animal urine and dung patches and from grazed legume-based pastures[1]

Gaseous loss	Urine Range measured	Urine Common range	Dung Range measured	Dung Common range	Grazed pasture Range measured	Grazed pasture Common range
(kg N ha⁻¹)						
N_2O emission	0–48	0–10	—	—	0.2–5	< 2
Total denitrification	0–73	5–25	2–80	10–30	3–17	3–6
Ammonia volatilization	37–170	50–110	2–156	25–75	1–17	5–10
(% N excreted)						
N_2O emission	0–14	0–7	—	—	0.3–2.0	< 1
Total denitrification	0–18	2–10	0.3–4	1–2	0.6–3.2	< 2
Ammonia volatilization	6–55	10–30	1–8	2–4	0.5–26	3–7

[1] Collated from data collected from temperate and tropical Australia, New Zealand, and the United Kingdom published by Wang et al. (1997), Fillery (2001), and Ledgard (2001).

sion rates reported from fertilized grasslands (Wang et al. 1997). Total annual losses of gaseous N from all sources in legume-based pastures commonly range from 1 to 25 kg N ha⁻¹ yr⁻¹ (Fillery 2001).

Many temperate pastures have an excess water and inorganic N over the winter to early spring period, and the progressive acidification of soils and high concentrations of nitrate detected in groundwater and streams are all evidence of nitrate leaching from pasture systems (Ridley et al. 2004). Estimates of nitrate leached below 1 m under annual clover/grass pastures receiving no fertilizer N range from 0 to more than 60 kg N ha⁻¹ yr⁻¹ (Ledgard 2001). The extent of nitrate leaching in any particular environment is regulated by the pattern and amount of rainfall leading to excess soil water in relation to evapotranspiration and by the soil's inherent capacity to hold water (Fillery 2001). Other variables that influence the accumulation of nitrate in soil will also affect the risk of nitrate leaching. Such factors include the presence of grazing animals (Whitehead 1995) and whether there are perennial species in the pasture or prolonged periods of fallow (Ridley et al. 2004).

Grazing animals also contribute to N inefficiencies by transferring excreta to nonproductive areas such as milking sheds, laneways, and gateways on dairy farms (Ledgard 2001) and "camps" in sheep-grazed pastures (Fillery 2001).

Contributions of Pasture Nitrogen to Crop Production

Before the rapid worldwide growth of the N fertilizer industry in the late 1940s, most crops were grown in rotation with pastures. Since then, continuous cropping has been

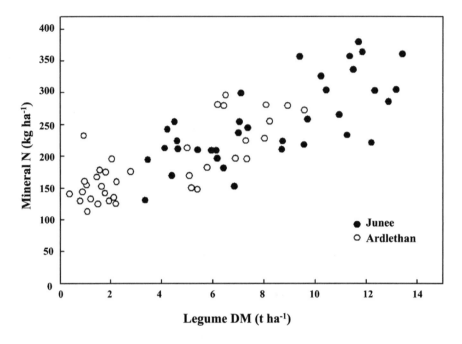

Figure 7.2. Relationship between concentrations of mineral N in the top 1 m of soil just before cropping and the total aboveground legume dry matter (DM) accumulated during the previous 3-year pasture phase (regression equation: mineral N = 130 + 0.0148 × legume DM, r^2 = 0.66). Data are derived from experiments undertaken on red clay loam (Rhodoxeralf) soils at two locations in southern New South Wales of Australia that differed in total average annual rainfall (550 mm at Junee, and 430 mm at Ardlethan; unpublished data from Virgona, Dear, Sandral and Swan).

increasingly adopted in most parts of the world, and pastures are increasingly being restricted to nonarable land. The major exceptions are in South America (particularly Brazil), Russia, republics of central Asia, and Australia. The main Australian crop, wheat, receives about 70 percent of its N requirement from mineralization of legume residues and soil organic matter and 30 percent from fertilizer (Angus 2001).

The return of above- and below-ground legume residues just before cropping can represent up to 200 to 300 kg of organic N ha^{-1} (Vinther and Jensen 2000), and it is commonly believed that the final pasture year has the largest impact on supplying mineral N to a following crop. Australian data, however, indicate that concentrations of mineral N accumulated following a pasture might be related more to the total amount of legume biomass grown over the entire pasture phase than to simply the year preceding cropping (Figure 7.2). The carryover of labile N from the pasture to the following crops can be as inorganic N mineralized during the end of the pasture phase

Table 7.3. Nitrogen uptake by barley (*Hordeum vulgare*), grain yield, and the proportion of grain nitrogen estimated to be derived from clover nitrogen following the incorporation of residues from either pure white clover or perennial ryegrass (*Lolium perenne*) swards or mixed white clover–ryegrass pastures[1]

| | Shoot | Grain | | |
| | N uptake | Yield | N offtake | % derived from |
Previous pasture	(kg N ha⁻¹)	(t ha⁻¹)	(kg N ha⁻¹)	clover N[2]
White clover	148	6.27	97	32
Clover-grass mix	128	6.07	87	24
Ryegrass	55	2.64	32	0

[1] Data represent the mean of two field experiments undertaken on a sandy loam (Typic Hapludalf) at Roskilde in Denmark (Jensen, unpublished).

[2] ¹⁵N-based estimates.

(Angus et al. 2000). It can also be as readily decomposed organic matter, leading to mineralization rates 30 percent greater in the first year after an annual pasture than in the second or subsequent years (Heenan and Chan 1992).

   Crop N uptake and grain yield following a pasture phase is generally related to the concentrations of soil mineral N present at the time of sowing in southeastern Australia (Angus et al. 2000). The accumulation of large amounts of inorganic N during a period of limited crop demand, however, does increase the risk of N losses through denitrification and leaching (Fillery 2001). Australian (Harris et al. 2002) and European (Hauggaard-Nielsen et al. 1998) studies have demonstrated that the amount and timing of N release from pasture residues and the resulting crop performance can be influenced by the proportions of clover and grass in the previous pasture sward. Some of the crop responses observed following clover-dominant swards may potentially reflect a reduced carryover of cereal roots disease compared with grassy pastures, but recent data confirm the direct contribution of N derived from clover residues to the crop (Table 7.3). This finding is consistent with the high concentrations of mineral N commonly detected immediately following legume-dominant pastures (e.g., see Figure 7.2). The rapid rates of mineralization imply that legume material breaks down much faster than the bulk of soil organic matter. An alternative explanation for the apparent accelerated net release of mineralized N following a grass-free pasture could be the result of less immobilization of N due to the absence of high C:N ratio grass residues (Peoples et al. 1998).

   Although the first crop sown after annual pastures can benefit directly from the flush of available N, the impact on soil mineral N and crop yield following annual pastures generally declines for the second crop in a cropping sequence and rarely persists into the third year (Harris et al. 2002). The N dynamics following Lucerne-based pastures differ from those of annual pastures. A perennial species such as Lucerne needs to be physically killed before returning to a cropping phase (Davies and Peoples 2003). Because most of the Lucerne-derived N is tied up in an organic form in the soil, the timing of the transition from Lucerne to cropping can be crucial in determining the release of mineral N and subsequent crop response (Angus et al. 2000). Mineralization immediately after Lucerne may be slower than in annual pastures for the following reasons: (1) soils tend to be much drier following Lucerne than after an annual pasture, and (2) there can be a transient period of N immobilization during decomposition of Lucerne residues (Bolger et al. 2003). The net result of a larger residual pool of legume N combined with a slower pattern of N mineralization is that crops grown after a vigorous Lucerne pasture tend to be supplied with N and yield better over a much longer period (3–4 years). This slow pattern of N release also has other potential advantages in that it should reduce the risk of a large pulse of mineralized N being leached below the cropping root zone before sowing.

The decision by farmers to supply crop N as fertilizer or as the by-product of previous pastures depends partly on the economics of continuous cropping and alternate phases of pasture and crops. The decline of pasture–crop rotations over the past half-century implies that the net returns from pastures, including their N contribution to the cropping phase, is perceived by farmers to be less than the net returns from continuous cropping. Another reason for the decline of pasture–crop rotations might be due to a discrepancy between crop N demand and the ability of pastures to supply sufficient N (Angus et al. 2000). This is partly because genetic improvements in the harvest index have driven crop yields, whereas N_2 fixation is tied to total biomass production, which is less amenable to genetic gain.

Conclusions

International emphasis on environmentally sustainable development based on the use of renewable resources is likely to refocus attention on the role of legumes to supply N for agriculture. This review gives an indication of the key factors that influence N inputs and losses from legume-based pastures, and it demonstrates the potential for such pastures to supply N for following crops. In non–N-fertilized, legume–pasture systems, N losses appear to be relatively low compared with highly fertilized grasslands or cropping systems. This appears to be related to the limited range in N inputs via N_2 fixation in part because of the self-regulation of N flows mediated through dynamic changes in the percent of Ndfa or shifts in pasture botanical composition in response to fluctuations in soil inorganic N and competition from associated grasses (Ledgard 2001).

Acknowledgments

The unpublished data presented in this paper were derived from research supported by either the Australian Grains Research and Development Corporation (GRDC) or the Danish Environmental Research Program.

Literature Cited

Angus, J. F. 2001. Nitrogen supply and demand in Australian agriculture. *Australian Journal of Experimental Agriculture* 41:277–288.

Angus, J. F., R. R. Gault, A. J. Good, A. B. Hart, T. J. Jones, and M. B. Peoples. 2000. Lucerne removal before a cropping phase. *Australian Journal of Agricultural Research* 51:877–890.

Bolger, T. P., J. F. Angus, and M. B. Peoples. 2003. Comparison of nitrogen mineralization patterns from root residues of *Trifolium subterraneum* and *Medicago sativa*. *Biology and Fertility of Soils* 38:296–300.

Davies, S. L., and M. B. Peoples. 2003. Identifying potential approaches to improve the reliability of terminating a lucerne pasture before cropping: A review. *Australian Journal of Experimental Agriculture* 43:429–447.

Dear, B. S., P. S. Cocks, M. B. Peoples, A. D. Swan, and A. B. Smith. 1999. Nitrogen fixation by subterranean clover (*Trifolium subterraneum* L.) growing in pure culture and in mixtures with varying densities of lucerne (*Medicago sativa* L.) or phalaris (*Phalaris aquatica* L.). *Australian Journal of Agricultural Research* 50:1047–1058.

De Neergaard, A., H. Hauggaard-Nielsen, L. S. Jensen, and J. Magid. 2002. Decomposition of white clover (*Trifolium repens*) and ryegrass (*Lolium perenne*) components: C and N dynamics simulated with the DAISY soil organic matter submodel. *European Journal of Agronomy* 16:43–55.

Fillery, I. R. P. 2001. The fate of biologically fixed nitrogen in legume-based dryland farming systems: A review. *Australian Journal of Experimental Agriculture* 41:361–381.

Giller, K. E. 2001. *Nitrogen fixation in tropical cropping systems,* 2nd ed. Wallingford, UK: CABI Publishing.

Harris, R. H., G. J. Scammell, W. Müller, and J. F. Angus. 2002. Wheat productivity after break crops and grass-free annual pasture. *Australian Journal of Agricultural Research* 53:1271–1283.

Hauggaard-Nielsen, H., A. de Neergaard, L. S. Jensen, H. Høgh-Jensen, and J. Magid. 1998. A field study of nitrogen dynamics and spring barley growth as affected by the quality of incorporated residues from white clover and ryegrass. *Plant and Soil* 203:91–101.

Heenan, D. P., and K. Y. Chan. 1992. The long-term effects of rotation, tillage and stubble management on soil nitrogen supply to wheat. *Australian Journal of Soil Research* 30:977–988.

Jarvis, S. C., D. Scholefield, and B. Pain. 1995. Nitrogen cycling in grazing systems. Pp. 381–419 in *Nitrogen fertilization in the environment,* edited by P. E. Bacon. New York: Marcel Dekker.

Jørgensen, F. V., E. S. Jensen, and J. K. Schjoerring. 1999. Dinitrogen fixation in white clover grown in pure stand and mixture with ryegrass estimated by the immobilized ^{15}N isotope dilution method. *Plant and Soil* 208:293–305.

Ledgard, S. F. 2001. Nitrogen cycling in low input legume-based agriculture, with emphasis on legume/grass pastures. *Plant and Soil* 228:43–59.

Ledgard, S. F., and K. W. Steele. 1992. Biological nitrogen fixation in mixed legume/grass pastures. *Plant and Soil* 141:137–153.

Peoples, M. B., and J. A. Baldock. 2001. Nitrogen dynamics of pastures: Nitrogen fixation inputs, the impact of legumes on soil nitrogen fertility, and the contributions of fixed N to Australian farming systems. *Australian Journal of Experimental Agriculture* 41:327–346.

Peoples, M. B., R. R. Gault, G. J. Scammell, B. S. Dear, J. Virgona, G. A. Sandral, J. Paul, E. C. Wolfe, and J. F. Angus. 1998. The effect of pasture management on the contributions of fixed N to the N-economy of ley-farming systems. *Australian Journal of Agricultural Research* 49:459–474.

Ridley, A. M, P. M. Mele, and C. R. Beverly. 2004. Developing sustainable legume-based farming systems in southern Australia. *Soil Biology and Biochemistry* (in press).

Thomas, R. J. 2000. Nitrogen fixation by forage legumes as a driving force behind the recuperation and improvement of soil quality in tropical agricultural systems: Opportunities for wider use of forage legumes? Pp. 539–540 in *Nitrogen fixation: From molecules to crop productivity*, edited by F. O. Pedrosa, M. Hungria, M. G. Yates, and W. E. Newton. Dordrecht: Kluwer Academic Publishers.

Thomas, R. J., and N. M. Asakawa. 1993. Decomposition of leaf litter from tropical forage grasses and legumes. *Soil Biology and Biochemistry* 25:1351–1361.

Thomas, R. J., N. M. Asakawa, M. A. Rondon, and H. F. Alarcon. 1997. Nitrogen fixation by three tropical forage legumes in an acid-soil savanna of Columbia. *Soil Biology and Biochemistry* 29:801–808.

Vinther, F. P. 1998. Biological nitrogen fixation in grass-clover affected by animal excreta. *Plant and Soil* 203:207–215.

Vinther, F. P., and E. S. Jensen. 2000. Estimating legume N_2 fixation in grass-clover mixtures of a grazed organic cropping system using two ^{15}N methods. *Agriculture, Ecosystems and Environment* 78:139–147.

Wang, Y.-P., C. P. Meyer, I. E. Galbally, and C. J. Smith. 1997. Comparisons of field measurements of carbon dioxide and nitrous oxide fluxes with model simulations for a legume pasture in southeast Australia. *Journal of Geophysical Research* 102:28013–28024.

Whitehead, D. C. 1995. *Grassland nitrogen*. Wallingford, UK: CABI Publishing.

8

Management of Nitrogen Fertilizer in Maize-based Systems in Subhumid Areas of Sub-Saharan Africa

B. Vanlauwe, N. Sanginga, K. Giller, and R. Merckx

The introduction of many scientific reports dealing with soil fertility management in sub-Saharan Africa (SSA), like the introduction of this chapter, starts with a statement referring either to the alarming negative and ever-declining nutrient balances at various scales or to the yield gap between potential and actual yields for major crops. Both observations are obviously strongly related. Smaling et al. (2002) reported N balances for SSA of -26 kg N ha^{-1} in 2000, compared with -20 kg N ha^{-1} in 1983. Tian et al. (1995), for instance, reported on-station yields of maize of 4000 kg ha^{-1} compared with on-farm yields of 1200 kg ha^{-1}.

The lack of investment in soil fertility regeneration and continuous exploitation of the natural capital has been identified as one of the root causes underlying the vicious cycle of low productivity–low income–low input use, leading to food insecurity for much of the rural population (CIAT et al. 2001). The Green Revolution had earlier changed farming in Asia and Latin America, but only minor achievements were made in SSA. Some of the reasons were the lack of the availability of new agricultural technologies in terms of improved crop varieties and crop management practices, the lack of existing farming systems able to support relatively intensive food production over long periods, and the lack of appropriate institutions. The lack of success in SSA and the environmental implications of the Green Revolution gradually moved fertilizer away from mainstream research and development agenda in favor of more organic-based farming systems.

Sub-Saharan Africa contains a diverse range of soil types, agroecologic zones, population densities, and market-access conditions. Results from the FAO Fertilizer Program showed an average response of 750 kg maize grain ha^{-1} to medium NPK applications with value-to-cost ratios for West African countries varying between 1.1 and 8.9 (FAO 1989). Responses to fertilizer application are likely to be less in areas with relatively high availability of land and good potential, however, for soil fertility-restoring fallow periods (e.g., Iganga District in Eastern Uganda), or in areas with inherently fertile soils (e.g., the Lama Depression in Benin and Togo; Vanlauwe et al. 2001a).

In this chapter, we focus on three areas where responses to N fertilizer are likely to be high because of high population densities and near absence of fallows. These areas are the Derived Savanna benchmark area in Southern Benin, more or less coinciding with the Mono Province (EPHTA 1996); the Northern Guinea Savanna benchmark area in Northern Nigeria, more or less coinciding with Kaduna State (EPHTA 1996); and Western Province in Kenya (Table 8.1). These three areas are also characterized by maize-based cropping systems and a subhumid climate.

Failure of Generic Fertilizer Recommendations in Sub-Saharan Africa

It is generally acknowledged that N fertilizer use is on average very low in SSA. What is often not acknowledged is the fact that when zooming in on fertilizer use at the community, farm, and field level, substantial variation exists in application rates. Once understood, this knowledge could be used to identify entry points for enhancing the use of N fertilizer in SSA.

Recommended and Current Use of Fertilizer in the Target Countries

In most countries in SSA, attempts were made to formulate fertilizer use recommendations for specific crops, valid for large areas. Several countries formulated recommendations at the (sub)national level (Table 8.1); others implemented major efforts to formulate recommendations at the district level, for example, in Kenya through the Fertilizer Use Recommendation Project (FURP) (Muriuki and Qureshi 2001). The current use, however, stands in sharp contrast with the recommended rates (Table 8.1).

The reasons behind the low fertilizer use in SSA are many. First, fertilizers are relatively expensive in SSA. In Western Kenya (Kitale), for instance, transporting fertilizer from the port of Mombassa nearly doubles the cost of one bag of diammonium phosphate (DAP) to about 17 U.S. dollars (USD) for a 50-kg bag (IFDC 2003). Notwithstanding the often dramatic responses to N fertilizer application that would easily pay back the investments in fertilizer applied, cash availability fluctuates during the year and financial resources may be limited when fertilizer needs to be bought. Second, N fertilizer is not always available in the correct formulation. In southern Benin, for instance, cotton fertilizer (14N-23P-14K-5S-1B) is usually available because of the presence of cotton in the farming systems, but this compound fertilizer obviously con-

Table 8.1. Selected characteristics of the target areas and recommended and current fertilizer use[1]

Characteristics	Southern Benin	Northern Nigeria	Western Kenya
Population density	200–700 persons km^{-2}	200–700 persons km^{-2}	400–1300 persons km^{-2}
Average farm size	1–5 ha	1–10 ha	0.5–2.5 ha
Livestock	Mainly small ruminants	High cattle density	Average cattle density
Cropping systems	Maize-based with cowpea, cotton	Maize-based with cowpea	Maize-based with beans
Soils	Ferralic nitisols and Lixisol–Acrisol–Leptosol association	Lixisols and Luvisols	Nitisols, Ferralsols, and Acrisols
Topography	Lowlands	Lowlands	Gently to very undulating (2–45%)
Agro-ecological zone	Subhumid (bimodal, about 1200 mm)	Subhumid (unimodal, about 900 mm)	Subhumid (bimodal, 1400–2200 mm)
Altitude	< 800 MASL	< 800 MASL	1200–1800 MASL
Recommended fertilizer rates at the national level[2]			
N	60 kg ha^{-1}	120 kg ha^{-1}	AEZ^3-dependent
P$_2$O$_5$	40 kg ha^{-1}	60 kg ha^{-1}	AEZ-dependent
K$_2$O	0 kg ha^{-1}	60 kg ha^{-1}	AEZ-dependent
Currently used fertilizer rate at the national level[2]			
N	2.3 kg ha^{-1}	1.5 kg ha^{-1}	4.9 kg ha^{-1}
P$_2$O$_5$	1.1 kg ha^{-1}	0.4 kg ha^{-1}	4.0 kg ha^{-1}
K$_2$O	1.3 kg ha^{-1}	0.4 kg ha^{-1}	0.5 kg ha^{-1}

[1] Sources: Benin: INRAB (1995); Kenya: Muriuki and Qureshi (2001); Nigeria: Balasubramanian et al. (1978).
[2] Assuming an average N, P$_2$O$_5$, and K$_2$O content of 30% of the commonly used fertilizers (http://www.fao.org/, 2001 data).
[3] AEZ, agro-ecozone; MASL, Mean Annual Sea Level.

tains too much P relative to N for application to maize. Third, knowledge of the appropriate and efficient use of N fertilizer is not equally widespread throughout SSA. Fourth, low or unstable produce prices may also limit farmers' interest in fertilizer use. Without trying to address fully the reasons leading to low N fertilizer use, the "organic" movement is also partly responsible for the negative aspects surrounding fertilizer use. Although organic resources certainly have a role to play in sustaining agricultural production in SSA, organic inputs alone are not going to supply all N required to boost crop production to acceptable levels (Place et al. 2003).

Spatial Differentiation of Fertilizer Use

Although aggregated N fertilizer application rates are low, a lot of variation exists in fertilizer use at various levels. Vanlauwe et al. (2002a) reported average N application rates of 8 kg N ha^{-1} yr^{-1} for Zouzouvou and 88 kg N ha^{-1} yr^{-1} for Eglimé, two villages within the Southern Benin benchmark area. Zouzouvou undergoes population-driven intensification and contains soils with lower inherent fertility compared with Eglimé, which undergoes market-driven intensification and consequently has much closer interactions with input and output markets (Manyong et al. 1996). In Western Kenya, Tittonell (2003) observed total mineral fertilizer inputs to vary between 0 kg ha^{-1} for the less densely populated Aludeka village in Teso District and 26 kg ha^{-1} for the densely populated Emuhaya village in Vihiga District for farmers with medium access to resources. For the wealthiest farmers, these values were 14 and 92 kg ha^{-1}, respectively.

Two other dimensions of variability that affect fertilizer use can be identified within villages in the target areas. First, farmers' resource endowment (e.g., access to cash for purchasing external inputs, access to irrigation facilities) strongly affects the use of fertilizer (Shepherd and Soule 1998), resulting in different fertilizer use between households within a village. In Shinyalu, the wealthier farmers apply 7 to 17 kg N ha^{-1}, whereas the poorest farmers apply only 0 to 3 kg N ha^{-1} (Tittonell 2003). Second, marked differences can often be observed in soil fertility status between fields within a farm. The variation in soil fertility status observed at the farm level (e.g., soil organic C varying from 0.2 to 2.2 percent from the bush fields to the homestead; Prudencio 1993) can be as high as the variation observed at the agroecozone level (e.g., soil C varying from 0.3 to 2.5 percent for soils from the equatorial forests to the Sudan savannas). Farmers are often aware of such gradients and use local terms to ascribe different soil quality features to different fields within their farm. This variation is caused by inherent soil properties, partly driven by their position in the landscape and by farmer-induced differences in management of the different fields. Tittonell (2003) observed fertilizer N inputs varying between less than 1 kg ha^{-1} for the remote fields and 4 kg ha^{-1} for the fields near the homestead for medium-wealth farmers in Shinyalu, Kakamega District, and Western Kenya.

Notwithstanding the existence of generic rules for N fertilizer use or recommendations, for most countries in SSA, application rates in the target countries are far below the recommended rates and vary according to farmer resource endowment and fertility status of the fields. These application rates are, however, mostly below the amount of N exported with harvested crops; consequently, the input rates and use efficiencies of fertilizer N need to be improved to arrest further soil fertility depletion. The following section gives evidence from the target areas on possible ways to achieve this improvement.

Fertilizer Use in an Integrated Soil Fertility Management Framework

Soil fertility management in the tropics has followed various research paradigms over the past decades. Following failures of earlier strategies to impact on rural livelihoods, the currently accepted paradigm is often described as integrated soil fertility management (ISFM) (Vanlauwe et al. 2002b). ISFM has been defined as "the development of adoptable and sustainable soil management practices that integrate the biological, chemical, physical, social, cultural and economic processes that regulate soil fertility" (CIAT et al. 2001). Technically, ISFM advocates the combined application of mineral and organic inputs and the optimization of their efficiency of use if it is compatible with the farmer's socioeconomic, cultural, and political environment.

Fertilizer Management

One of the most straightforward approaches to improving fertilizer use efficiency is by managing the type of fertilizer and the mode of application. Without giving an exhaustive review on this topic, a number of points can be reiterated. Mughogho et al. (1986) reported that for subhumid West Africa, the source of fertilizer N (CAN v Urea), had little effect on plant recovery of N. Also, differences between broadcasting and banding the fertilizer did not yield significant differences in N recovery, although both methods of application resulted in higher recoveries than point-placed urea super-granules. This is in contrast with the current recommendations for N fertilizer management (deep placement, spot applied) advocated by the Sasakawa Global 2000 program in Northern Nigeria (Iwuafor et al. 2002). Usually, rainfall in the subhumid areas is sufficient to avoid extended exposure of urea on the soil surface and consequent losses through volatilization (Arora et al. 1987). Application of N fertilizer is also known to induce soil acidification, the degree depending on the chemical composition of the fertilizer (Vanlauwe et al. 2001a). This is particularly important in areas with low soil-buffering capacities. In the target areas, N fertilizer is usually split applied to avoid excessive leaching losses. N is usually applied together with P at planting (e.g., cotton fertil-

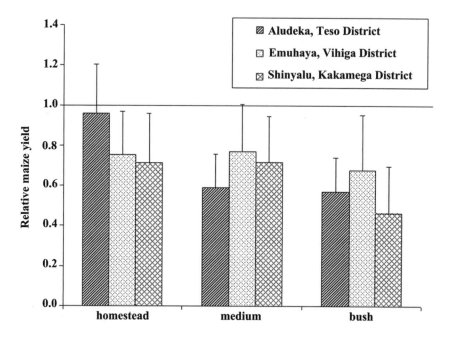

Figure 8.1. Relative maize yields (yield in the treatment with P and K applied over yield in the treatment with N, P, and K applied) for homestead fields and fields at medium and far away distances from the homestead in Teso, Vihiga, and Kakamega Districts, Western Kenya. Data are average values of 6 limiting nutrient strips. Error bars are standard deviations.

izer in Benin, NPK in Nigeria, DAP in Kenya), and urea is commonly used for top dressing.

Targeting Niches

As highlighted in the previous section, the various production units in a single farm can show great differences in soil fertility status, and this is likely going to influence crop production and use efficiency of fertilizer N. Carsky et al. (1998) reported a positive relationship between unfertilized maize yields and the soil organic C content for a number of sites in Northern Nigeria. In Western Kenya, the relative yield of maize in the absence of N was higher for the homestead fields than for the bush fields in Teso and Kakamega Districts (Figure 8.1). In Vihiga District, where farm sizes are very small, relative yields were more consistent across the various field types (Figure 8.1).

An interesting research issue is whether the returns to N fertilizer application are higher on soils with a high soil fertility status, such as the homestead fields, compared with soils with lower soil fertility status. Soil organic matter (SOM) contents are usu-

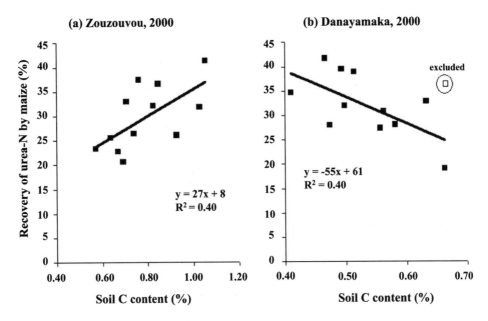

Figure 8.2. Observed relationships between recovery of ^{15}N labeled urea N in the maize shoot biomass and the soil organic C content for 12 farmers' fields in Zouzouvou (Southern Benin) and Danayamaka (Northern Nigeria). Urea was split-applied (one third at planting, two-thirds at knee height) at 90 kg N ha[-1] in Zouzouvou and 120 kg N ha[-1] in Danayamaka. One observation was excluded from the regression analysis for the Danayamaka data.

ally positively related with specific soil properties or processes fostering crop growth, such as cation-exchange capacity, rainfall infiltration, or soil structure. In plots where any of these constraints limit crop growth, a higher SOM content may enhance the demand by the crop for N and consequently increase the fertilizer N use efficiency. On the other hand, SOM also releases available N that may be better synchronized with the demand for N by the plant than fertilizer N, and consequently a larger SOM pool may result in lower use efficiencies of the fertilizer N. A preliminary investigation using ^{15}N labeled urea under on-farm conditions in Southern Benin and Northern Nigeria revealed contrasting trends between the two sites (Figure 8.2). Although the exact reasons underlying the different trends presented are not clear, the major function of the SOM pool in Benin is likely mainly to alleviate one or more specific constraints to crop growth besides N while in Nigeria, SOM mainly supplies N to the growing crop.

As mentioned, farmers usually appreciate these differences in soil fertility status between production units and manage these units differently. Local terms are usually used to describe the different units, and these local appreciations of soil fertility status usually correlate very well with formal assessments of soil fertility status (Tittonell 2003).

Organic–Mineral Interactions

ISFM advocated the combined application of mineral and organic inputs because (1) either of the two inputs is usually not available in sufficient quantities, (2) both inputs are needed in the long run to sustain soil fertility and crop production, and (3) positive interactions between both inputs could potentially result in added benefits in terms of extra grain yield or extra soil fertility increase. Direct and indirect *hypotheses* were devised by Vanlauwe et al. (2001a) to explain the occurrence of such benefits. The direct hypothesis was formulated as follows: "Temporary immobilization of applied fertilizer N may improve the synchrony between the supply of and demand for N and reduce losses to the environment." The indirect hypothesis was formulated for N supplied as fertilizer as "any organic matter-related improvement in soil conditions affecting plant growth (except N) may lead to better plant growth and consequently enhanced efficiency of the applied N." The indirect hypothesis recognizes that organic resources can have multiple benefits besides the short-term supply of available N. Such benefits could be an improved soil P status by reducing the soil P sorption capacity, improved soil moisture conditions, less pest and disease pressure in legume-cereal rotations, or other mechanisms. Both hypotheses, when proven, lead to an enhancement in N use efficiency, through improvement of the N supply *(direct hypothesis)* and the demand for N *(indirect hypothesis)*. Obviously, mechanisms supporting both hypotheses may occur simultaneously.

Testing the direct hypothesis with ^{15}N labeled fertilizer, Vanlauwe et al. (2001a) concluded that direct interactions between organic matter (OM) and fertilizer N can be demonstrated under field conditions. These interactions were affected by resource quality and the method of incorporation of the applied organic resources. In a multi-locational trial with external inputs of organic matter, Vanlauwe et al. (2001b) observed added benefits from the combined treatments at two of the four sites, which experienced serious moisture stress during the early phases of grain filling. The positive interaction at these two sites was attributed to the reduced moisture stress in the "mixed" treatments compared with the sole urea treatments because of the presence of organic materials (surface and subsurface placed) and constitutes evidence for the occurrence of mechanisms supporting the indirect hypothesis. Although more examples can be found in literature supporting the indirect hypothesis, it is clear that a wide range of mechanisms could lead to an improved use efficiency of applied external inputs. These mechanisms may also be site specific; for example, an improvement in soil moisture conditions may be less relevant in a humid forest environment. Unraveling these as a function of easily quantifiable soil characteristics is a major challenge and needs to be done to optimize the efficiency of external inputs. On the other hand, when applying organic resources and mineral fertilizer simultaneously, one hardly ever observes negative interactions, indicating that even without clearly understanding the mechanisms underlying positive interactions, applying organic resources in combination with mineral inputs stands as an appropriate fertility management principle.

Resilient Germplasm

Besides managing the supply of available N to the crops, N fertilizer use can also be improved by enhancing the demand for N through use of N efficient germplasm, such as the Oba super 2 maize variety, currently the most widely grown maize hybrid in Nigeria (Sanginga et al. 2003). These varieties produce significantly higher yields than the traditional varieties at low N and respond at least as well as the traditional varieties to N fertilizer application. A new open pollinated maize variety with similar characteristics is now also ready for release to farmers. Maize varieties resistant to *Striga* and tolerant to drought have also been developed that could, through their improved resilience against unfavorable growth conditions, trigger better utilization of applied fertilizer N.

Getting the Message Across: Modeling and Decision Aids for Fertilizer Use in Smallholder Agriculture in Sub-Saharan Africa

A wealth of knowledge has been accumulated on different approaches to manage soil fertility within smallholder farms, but uptake of apparently promising technologies and measures by farmers is limited. The fundamental problem is often the assumption that technologies that increase yields at plot scale will be rapidly adopted by farmers. Adoption in fact depends on how well innovations fit into the whole farm livelihood, which is determined not only by yield improvement, but also by competition with other activities for resources, in particular land and labor; by the development of local markets or links to distant markets for product and input factors; and by institutional support for learning about and adapting the innovation. To understand the complex interactions between socioeconomic and agroecologic factors that are variable in both time and space and to analyze how future food security and management of environmental services can be improved, models that incorporate the dynamic interactions between these factors are essential. One such approach that is currently being developed is the NUANCES (Nutrient Use in Animal and Cropping systems—Efficiency and Scales) framework (Giller et al. 2003). Such a framework allows advising on how best to use the limited amount of N fertilizer available to the farmers, keeping in mind the variability in soil fertility status within the farm, the functioning of the input and output markets, and the farmers' resource endowment.

Relevant information needs to be synthesized in a quantitative framework, and that framework needs to be translated in a format accessible to the end users. The level of accuracy of such a quantitative framework is an important point to consider. The generation of a set of rules of thumb is likely to be more feasible than software-based aids that generate predictive information for a large set of environments. The level of com-

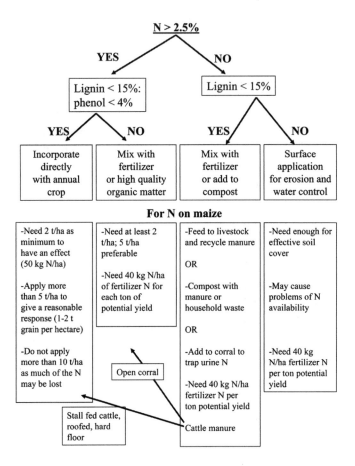

Figure 8.3. A potential framework for adding quantitative information regarding N management for a maize crop to the conceptual decision support system for organic N management, developed by Palm et al. (2001). The decision support system for organic N management that proposes optimal ways to manage organic resources depending on their quality, expressed as their N, lignin, and soluble polyphenol content. Source: Giller (2001).

plexity is another essential point to take into consideration. For instance, if variation between fields within one farm is large and affects ISFM practices, this may justify having this factor included in decision aids. Giller (2001), for instance, proposed "rule of thumb" values for fertilizer N application rates in combination with organic resources of varying quality, based on the decision support system for organic N management, conceptualized by Palm et al. (2001) (Figure 8.3).

Other aspects that will influence the way information and knowledge are condensed into a workable package are (1) the targeted end-user community, (2) the level of specificity required by the decisions to be supported, and (3) level of understanding generated related to the technologies targeted. Van Noordwijk et al. (2001) prefer the term *negotiation support systems* because the term *decision support systems* suggests that a single authority makes decisions that will then be imposed on the various stakeholders. In an ISFM context, it is recognized that different stakeholders may have conflicting interests related to certain specific soil management strategies and that a certain level of negotiation may be required.

Literature Cited

Arora, Y., L. A. Nnadi, and A. S. R. Juo. 1987. Nitrogen efficiency of urea and calcium ammonium nitrate for maize (*Zea mays*) in humid and subhumid regions of Nigeria. *Journal of Agricultural Science* 109:47–51.

Balasubramanian, V., L. A. Nnadi, and A.U. Mokwunye. 1978. Fertilizing sole crop maize for high yields. *Samaru Miscellaneous Papers* No. 76. Institute of Agricultural Research, Ahmadu Bello University, Zaria, Nigeria, 14 pp.

Carsky, R. J., S. Jagtap, G. Tian, N. Sanginga, and B. Vanlauwe. 1998. Maintenance of soil organic matter and N supply in the moist savanna zone of West Africa. Pp. 223–236 in *Soil quality and agricultural sustainability*, edited by R. Lal. Chelsea, Michigan: Ann Arbor Press.

CIAT, TSBF, and ICRAF (Central Internacional de Agricultura Topical, Tropical Soil Biology and Fertility Program, and International Centre for Research in Agroforestry). 2001. *Soil fertility degradation in sub-Saharan Africa: Leveraging lasting solutions to a long-term problem.* http://www.ciat.cgiar.org/tsbf_institute/index_tsbf.htm.

EPHTA. 1996. *Mechanism for sustainability and partnership in agriculture.* Ibadan, Nigeria: International Institute for Tropical Agriculture.

FAO (Food and Agricultural Organization of the United Nations). 1989. *Fertilizers and food production. The FAO fertilizer programme 1961–1986.* Rome, Italy: FAO.

Giller, K. E. 2001. Organic-inorganic nutrient sources and interactions. Pp. 19–22 in *Integration of soil research activities in eastern and southern Africa*, edited by C. Palm, A. Bationo, and S. Waddington. Nairobi, Kenya: Tropical Soil Biology and Fertility Program.

Giller, K. E., E. Rowe, N. de Ridder, and H. van Keulen. 2004. Resource use dynamics and interactions in the tropics: Moving to the scale of the "livelihood." *Agricultural Systems* (in press).

IFDC (International Fertilizer Development Center). 2003. *An assessment of fertilizer prices in Kenya and Uganda: Domestic prices vis-à-vis international market prices.* IFDC Paper Series IFDC-PCD-27. Muscle Shoals, Alabama: IFDC.

INRAB. Institut National de Recherches Agricoles du Benin. 1995. *Fiche technique. Cultures vivrieres. Cereules, legumineuses a graines et tubercules.* INRAB, Cotonou, Republique du Benin.

Iwuafor, E. N. O., K. Aihou, B. Vanlauwe, J. Diels, N. Sanginga, O. Lyasse, J. Deckers, and R. Merckx. 2002. On-farm evaluation of the contribution of sole and mixed applications of organic matter and urea to maize grain production in the savanna. Pp. 185–

197 in *Integrated plant nutrient management in sub-Saharan Africa: From concept to practice*, edited by B. Vanlauwe, J. Diels, N. Sanginga, and R. Merckx. Wallingford, UK: CABI Publishing.

Manyong, V. M., J. Smith, G. K. Weber, S. S. Jagtap, and B. Oyewole. 1996. *Macrocharcterization of agricultural systems in West Africa: An overview*. Ibadan, Nigeria: International Institute for Tropical Agriculture.

Mughogho, S. K., A. Bationo, and B. Christianson. 1986. Management of nitrogen fertilizers for tropical African soils. Pp. 117–172 in *Management of nitrogen and phosphorus fertilizers in sub-Saharan Africa*, edited by P. L. G. Vlek and A. U. Mokwunye. Dordrecht, The Netherlands: Martinus Nijhoff Publishers.

Muriuki, A. W., and J. N. Qureshi. 2001. *Fertilizer use manual: A comprehensive guide on fertilizer use in Kenya*. Nairobi, Kenya: Kenya Agricultural Research Institute.

Palm, C. A., C. N. Gachengo, R. J. Delve, G. Cadisch, and K. E. Giller. 2001. Organic inputs for soil fertility management in tropical agroecosystems: Application of an organic resource database. *Agriculture, Ecosystems and Environment* 83:27–42.

Place, F., C. B. Barrett, H. A. Freeman, J. J. Ramisch, and B. Vanlauwe. 2003. Prospects for integrated soil fertility management using organic and inorganic inputs: Evidence from smallholder African agricultural systems. *Food Policy* 28:365–378.

Prudencio, C. Y. 1993. Ring management of soils and crops in the West African semi-arid tropics: The case of the mossi farming system in Burkina Faso. *Agriculture, Ecosystems and Environment* 47:237–264.

Sanginga, N., K. Dashiell, J. Diels, B Vanlauwe O. Lyasse, R. J. Carsky, S. Tarawali, B. Asafo-Adjei, A. Menkir, S. Schulz, B. B. Singh, D. Chikoye, D. Keatinge, and R. Ortiz. 2003. Sustainable resource management coupled to resilient germplasm to provide new intensive cereal—grain legume—livestock systems in the dry savanna. *Agriculture, Ecosystems and Environment* 100:305–314.

Shepherd, K., and M. J. Soule. 1998. Soil fertility management in west Kenya—dynamic simulation of productivity, profitability and sustainability at different resource endowment levels. *Agriculture, Ecosystems and Environment* 71:133–147.

Smaling, E. M. A., J. J. Stoorvogel, and A. de Jager. 2002. Decision making on integrated nutrient management through the eyes of the scientist, the land-user and the policy maker. Pp. 265–284 in *Integrated plant nutrient management in sub-Saharan Africa: From concept to practice*, edited by B. Vanlauwe, J. Diels, N. Sanginga, and R. Merckx. Wallingford, UK: CABI Publishing.

Tian, G., B. T. Kang, I. A. Akobundu, and V. M. Manyong. 1995. Food production in the moist savanna of West and Central Africa. Pp. 107–127 in *Moist savannas of Africa: Potentials and constraints for crop production*, edited by B. T. Kang, I. O. Akobundu, V. M. Manyong, R. J. Carsky, N. Sanginga, and E. A. Kueneman. Ibadan, Nigeria: International Institute for Tropical Agriculture.

Tittonell, P. 2003. *Soil fertility gradients in smallholder farms of western Kenya: Their origin, magnitude and importance. Quantitative approaches in systems analysis No. 25*. Wageningen, The Netherlands: The C.T. de Wit Graduate School for Production Ecology and Resource Conservation.

Vanlauwe, B., J. Wendt, and J. Diels 2001a. Combined application of organic matter and fertilizer. Pp. 247–280 in *Sustaining soil fertility in West-Africa*, edited by G. Tian, F. Ishida, and J. D. H. Keatinge. Special Publication No. 58, Madison, Wisconsin: Soil Science Society of America.

Vanlauwe, B., K. Aihou, S. Aman, E. N. O. Iwuafor, B. K. Tossah, J. Diels, N. Sanginga, R. Merckx, and S. Deckers. 2001b. Maize yield as affected by organic inputs and urea in the West-African moist savanna. *Agronomy Journal* 93:1191–1199.

Vanlauwe, B., J. Diels, O. Lyasse, K. Aihou, E. N. O. Iwuafor, N. Sanginga, R. Merckx, and J. Deckers. 2002a. Fertility status of soils of the derived savanna and northern guinea savanna and response to major plant nutrients, as influenced by soil type and land use management. *Nutrient Cycling in Agroecosystems* 62:139–150.

Vanlauwe, B., J. Diels, N. Sanginga, and R. Merckx. 2002b. *Integrated plant nutrient management in sub-Saharan Africa: From concept to practice.* Wallingford, UK: CABI Publishing.

Van Noordwijk, M., M. T. P Toomich, and B. Verbist. 2001. Negotiation support models for integrated natural resource management in tropical forest margins. *Conservation Ecology* 5:21–28.

9

Integrated Nitrogen Input Systems in Denmark

J. E. Olesen, P. Sørensen, I. K. Thomsen, J. Eriksen, A. G. Thomsen, and J. Berntsen

Cycling of N in agriculture through the use of mineral fertilizers, manures, and N-fixing crops gives rise to many forms of N emissions to the environment, including nitrate (NO_3) leaching, ammonia (NH_3) volatilization, and nitrous oxide (N_2O) emissions, resulting in groundwater pollution, eutrophication of surface waters, soil acidification, and contributions to global warming.

The high rates of N input in intensive North European agricultural systems have given rise to high loss rates, and the focus in Danish agriculture during the past two decades has been on increasing the N use efficiency (NUE) with the aim of reducing losses. The NUE at the system level can be increased by improved handling of manure, targeted application of fertilizers and manures, and through adjustments of the crop rotation.

Trends in Danish Agriculture

The agricultural area in Denmark constituted 62 percent (26,470 km^2) of the total land area in 2000. Grasslands constituted 30 percent and cereals 47%, with dairy cattle and pigs dominating livestock production. The cattle population declined by 34 percent, from 2.84 million in 1970 to 1.87 million in 2000; however, milk was almost constant as a consequence of increasing productivity. The pig population increased by 43 percent from 8.36 million in 1970 to 11.92 million in 2000. Most of the animal feed is produced domestically.

During the 1990s fertilizer N declined rapidly (Figure 9.1), resulting in an increase in N recovery efficiency (RE) over the 20-year period of 12 percent based on total N input and of 18 percent based on input of manure, waste, and mineral N fertilizer only.

The change in mineral fertilizer use has been a result of the Danish Action Plan on the Aquatic Environment, which was initiated in 1987 and aimed at reducing N leach-

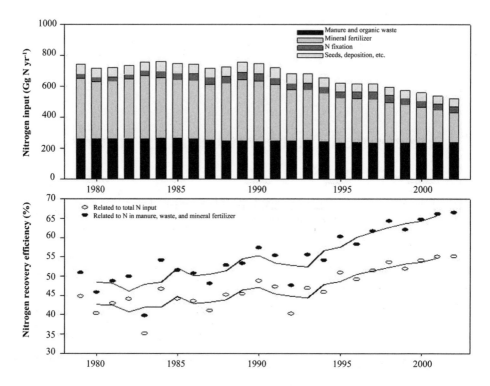

Figure 9.1. Annual N input to fields (top graph) and nitrogen use efficiency estimated as harvested N in proportion of either total N input or N in manure, organic waste, and mineral fertilizer only (bottom graph) (Kyllingsbæk, 2000).

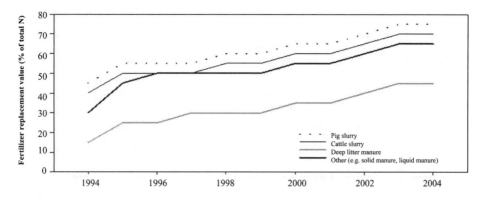

Figure 9.2. Change in fertilizer replacement value of different manure types in the Danish farm-scale fertilizer accounting system.

ing from rural areas by 100,000 t N yr^{-1} (Grant et al., 2000). The measures in the Aquatic Action Plan can be grouped into four main categories: (1) improved use of animal manures, (2) reduced N input, (3) improved crop rotations (including cover crops), and (4) conversion of cropland to permanent grassland or forestry. Fertilizer planning and nutrient accounting are compulsory with a limit on allowable use of N in fertilizers and manures, which has been set at 10 percent below the economical optimal N rate. The required fertilizer replacement value for manures has been gradually increased over time (Figure 9.2), depending on the manure type.

Crop Nitrogen Demand

The crop N demand generally increases with increasing yield. The N demand is affected by soil and climate conditions but also by biotic conditions, such as the occurrence of weeds and diseases. Olesen et al. (2003) showed that the estimated optimal N fertilizer rate for untreated diseased winter wheat was 60 kg N ha^{-1} lower than for crops without disease. The use of fungicides with an efficacy twice that of the particular fungicide type used in the experiment would increase the optimal N rate by about 20 kg N ha^{-1}.

Determining the Crop Nitrogen Demand at Field Level

The software system BEDRIFTSLØSNING is used by most Danish farmers to estimate crop N demand at field level. A simple soil organic matter model is used to estimate the residual effects of previous crops, manure application, and other inputs of organic matter on net soil N mineralization. BEDRIFTSLØSNING is supplied with information on crops grown in the previous 2 years and the expected yield. By subtracting N mineralized and N supplied with animal manure in the actual growing season, the recommended application rate of mineral fertilizer is calculated. The system has been successful in predicting the overall optimal N rate at farm level, but the skill at field level is still rather low.

To improve the estimates of optimum N rate, three biological/chemical methods for determining potentially mineralizable soil N (anaerobic incubation, boiling with KCl, and chloroform fumigation) were tested in Denmark in 2000 and 2001 (Thomsen et al. 2003). The amount of N mineralized by the three methods was compared with the actual crop N uptake in the field. Of the three methods, the anaerobic incubation gave the best correlation between crop N uptake and mineralized N; however, less than 40 percent of the variation in crop N uptake could be explained by the results from the anaerobic incubation.

Determining the Crop N Demand at Sub-Field Level

During 2001 to 2003, field experiments were conducted with the aim of developing algorithms for the redistribution of N based on soil and plant sensors (Broge et al. 2003). Each year plots placed in fields at different sites in Denmark received 60, 120, 180, or

240 kg N ha^{-1} in two dressings. Measurements with plant sensors (ratio vegetation index [RVI]) were made just before the second application, while soil sensors (electrical conductivity) were used either before planting or after harvest. A statistical analysis of the relationship between N-rate, plot yield, and sensor measurements showed a significant relationship for nearly all fields between sensor measurements and optimum N application rate. The relationship was applied for redistributing a certain amount of N fertilizer on a field, such that N fertilizer was moved from areas with low and high RVI values (low and high amounts of biomass) to areas with medium RVI. Regardless of the location, year, and sensor combination, the yield benefit from this redistribution was small and averaged less than 10 kg grain ha^{-1}. Thus the economic gain that can be expected from redistributing N within fields is nearly nil (Berntsen et al. 2002), and the redistribution of an a priori fixed amount of N within fields does not seem to have much prospect in Denmark. The current work on crop sensors and algorithms therefore focuses on monitoring the absolute N status of the crops and need for additional N.

Efficient Use of Manures

First-year Effects and Residual Effects

Most manures are stored under anaerobic or partly anaerobic conditions, and after application to soil, part of the ammonium N is immobilized by soil microorganisms as a result of the presence of easily decomposable compounds in the manure (Kirchmann and Lundvall 1993). Organic N in the manure is also mineralized, and after 2 to 3 months, N mineralization is often equal to N immobilization. Thus, the potential first year N effect of most manures is equivalent to the ammonium content of the manure (Jensen et al. 1999). High N utilization can be achieved only if losses of N by leaching, denitrification, and volatilization are minimized.

In a number of Danish experiments, the availability of manure N was measured in small confined plots using ^{15}N-labeled feces and urine from animals fed on ^{15}N-labeled diets. The enclosures used and the applied nutrient application rates allowed normal plant growth. By using labeled and similar unlabeled materials, the contribution from feces, urine, and bedding material to crop N uptake was determined separately. The crop uptake of labeled N in the year of application is highly influenced by the origin of the labeled N (Table 9.1). A significant part of the manure N is still in the soil after harvest of the first crop and is released slowly resulting in both losses and residual N effects during the following years. During the autumn/winter period following manure application to spring barley, 2 to 5 percent of the applied labeled N from both animal manure and mineral fertilizer was lost by nitrate leaching from bare soil after barley harvest (Thomsen et al. 1997). When barley was undersown with a ryegrass cover crop, 1.5 to 6 percent of the labeled manure N was recovered in the cover crop (Jensen et al. 1999; Sørensen et al. 1994; Sørensen and Jensen, 1998).

Table 9.1. Crop uptake of ^{15}N-labeled mineral fertilizer and animal manure components during two or three growing seasons measured in Danish experiments under field conditions

| ^{15}N-labeled component | ^{15}N crop uptake (% of input) | | | |
	Appl. year	1st year[1]	2nd year	References[2]
Applied in spring before sowing				
spring barley				
Mineral fertilizer N	36–57	3–5	1.2–1.5	1–4
Ruminant feces in slurry	12–17	3–6	—	1,2
Ruminant feces in solid manure	9	4.1	1.1–2.0	3
Pig feces in slurry	33	4.2	—	5
Ruminant urine in slurry	32-36	3	—	2
Ruminant urine in solid manure	25–27	3.6	1.3	3
Pig urine in slurry	47	2.5	—	5
NH_4-N in pig and cattle slurry	27–41	3–4	1.8–2.5	4, 5
Bedding straw in solid manure	9–10	3.3	1.1–1.3	3
Applied in August before				
winter wheat				
Total N in solid manure	8–10	2.6	—	6

[1] First year of residual effects.

[2] 1: Sørensen et al. (1994), 2: Thomsen et al. (1997), 3: Jensen et al. (1999), 4: Sørensen and Amato (2002), 5: Sørensen, unpublished, 6: Thomsen (2001).

In the second year (first residual year), 3 to 6 percent of the labeled N was recovered in barley and grass crops, and in the third year another 1 to 2.5 percent was recovered (Table 9.1). A few months after application, the release rate of residual labeled N is lower for fecal N and straw N than for urinary and mineral fertilizer N (calculated as percentage of residual ^{15}N in soil), but because more N is left in the soil from feces and straw than from urine and fertilizer, the ^{15}N release, calculated as the percentage of applied ^{15}N, is similar for N in the different components (Table 9.1). Manure storage conditions have negligible influence on the release of residual N (Thomsen 2001). Thus, the residual N effect in the years after application of animal manure is mainly determined by the amount of total N applied (manure + mineral fertilizer), not the type of manure.

Table 9.2 shows average estimates of residual N effects after a single and repeated application of standard animal manure types. The estimates are based on ^{15}N and other experiments and show the additional effect of animal manure N compared with soil receiving mineral fertilizer N (Sørensen et al. 2002). When comparing residual N effects in such experiments, it should be recognized that there is also crop uptake of residual fertilizer ^{15}N in the years after mineral fertilizer application. The residual effect

Table 9.2. Estimated residual nitrogen effects of repeated applications of animal manure supplemented with mineral fertilizer compared with soil receiving only mineral fertilizers, expressed as fertilizer replacement value (% of annual manure application)[1]

	Repeated animal manure applications		
Manure type	1 yr [2]	2 yr	10 yr
Cattle slurry	3–5	5–7	10–15
Pig slurry	2–4	3–5	7–10
Solid manure	6–9	8–12	16–24

[1] From Sørensen et al. (2002).
[2] First year with residual N effect.

in the first year after manure application is relatively low, whereas the effect of repeated manure application can be considerable (Table 9.2). A large part of the residual manure N is mineralized during autumn (Sørensen and Amato 2002), and residual effects are therefore higher in crops with a long growing period.

Effects of Manure Treatment, Application Time, and Method

Animal manure can be treated in different ways to modify its characteristics, for example, slurry separation, anaerobic digestion, and slurry acidification. To increase the first-year availability of manure N, it is necessary to remove part of the decomposable carbon in the manure without losing N during the process. After anaerobic digestion of slurry, the content of decomposable carbon is reduced and part of the organic manure N is mineralized, resulting in less N immobilization after application and a higher plant availability of N (Kirchmann and Lundvall 1993).

Trail-hose application (surface-banding) in spring in established cereal crops is widely used in Denmark. A well-established crop canopy reduces ammonia volatilization, but significant volatilization may still occur when slurry is surface-banded in a crop (Sommer et al. 1997). The ammonia emission can be further reduced by direct injection, but crop damage by injector tines and more traffic in the field by injection is a problem in established crops, and can reduce yields.

A high N utilization is obtainable after direct injection of slurry before sowing in spring. The high N utilization is partly due to reduced volatilization and partly due to lower N immobilization when the slurry is placed in a band in the soil (Sørensen and Jensen 1998). Sørensen et al. (2003) found higher denitrification losses after direct injection, but this loss was counterbalanced by the lower N volatilization loss. If slurry injection is followed by wet soil conditions the loss of N by denitrification can be significant (Thompson et al. 1987).

Efficient Use of Nitrogen in Crop Rotations

Effects of Previous Crops

Fertilizer N standards are enforced for farmers and growers in Denmark. The standards are field and crop specific and based on average N response curves from field trials carried out primarily on farmers' fields. The fertilizer standards are made for individual crops and are made as specific as possible for different soil types and previous crops. The largest experimental base is available for winter wheat, and this has made it possible to distinguish between different previous crops.

An analysis of the data for winter wheat shows that the economic optimum N application rate may vary considerably within years and sites (Petersen 2002). The optimum N rate ranged from 0 to 300 kg N ha^{-1} within a single year. One of the important aspects was the previous crop grown on a field, with grass–clover and alfalfa having a residual effect that reduced optimum N rate by 60 kg N ha^{-1} (Petersen 2002). Even the crop grown 2 years previously had an effect if the crop was grass–clover or alfalfa, where optimum N rate was reduced by 50 kg N ha^{-1}. The effect of winter rape as the previous crop was a reduction in optimum N rate of about 45 kg N ha^{-1} and that of spring rape and pulse crops was a reduction of about 35 kg N ha^{-1}.

Residual Effects of Grasslands

In grasslands, a considerable buildup of N may take place. As a consequence of this buildup, the cultivation of grasslands is followed by a rapid and extended period of N mineralization that may often exceed the requirement of the subsequent arable crop (Francis 1995). The residual effects of six 3-year-old grasslands on yield and nitrate leaching in the following three cereal crops were investigated on a loamy sand in Denmark (Eriksen 2001). The grasslands were unfertilized grass–clover and fertilized ryegrass subject to cutting or continuous grazing by dairy cows with two levels of N in feed supplements. In the first year the residual effect of the grazed grasslands was sufficient to obviate the need for supplementary fertilizer, but in the following years gradually more fertilizer N was required to obtain economic optimal yields. A residual effect (N fertilizer replacement value) following grass–clover was at least 115 kg N ha^{-1}. The residual effect of grazed ryegrass was 90 to 100 kg N ha^{-1}, while for cut ryegrass it was only 25 kg N ha^{-1}. In the second year after grassland cultivation, the residual effects were 60 kg N ha^{-1} after grass–clover, 40 kg N ha^{-1} after grazed ryegrass, and negative after cut ryegrass. In the third year, the residual effects were either very small or nonexistent.

The residual effect is a combination of non-N and N effects. The N recovery of grass–clover and grazed and cut ryegrass in the first year was 50 to 70, 40 to 50, and less than 0 kg N ha^{-1}, respectively.

Effect of Cover Crops

Cover crops are grown between main crops with the purpose of reducing nitrate leaching from soil during autumn and winter when the soil would have been bare. The commonly tested cover crop in Denmark is perennial ryegrass (*Lolium perenne* L.) under-sown in spring barley. The N uptake may vary with soil type and N application rates in the specific field, but the average N uptake in a ryegrass cover crop grown after spring barley is about 24 kg N ha^{-1} (Hansen et al. 2000c). Uptake of N from ryegrass material incorporated in autumn ranges from 10 to 19 percent, 2 to 4 percent, and 1 to 2 percent in the first, second, and third year, respectively (Jensen 1992; Thomsen and Jensen 1994). N derived from the cover crop may accordingly amount to 2 to 5 kg N ha^{-1} in the first growth season after incorporation. If the N taken up by the cover crop would not have been completely lost by leaching during winter, however, the cover crop, through its uptake of soil N, may effectively reduce the soil N pool available for initial uptake of the subsequent crop (Thorup-Kristensen 1993). Because a ryegrass cover crop may contain 3 to 9 kg N ha^{-1} at the time of harvest of the main crop (Jensen 1991), an immediate positive residual value may not be obtained if determined in barley grown repeatedly with a cover crop.

The N content of soil is raised after long-term use of cover crops (Thomsen 1995) and thereby also the amount of mineralizable N. Extra N mineralized from cover crops may either be taken up by a crop or lost by nitrate leaching. After 24 years of repeated use of cover crops, Hansen et al. (2000a) found that the average increase in leaching over 4 years corresponded to 14 kg N ha^{-1} yr^{-1}; however, the amount of plant-available N was also increased, thereby reducing the need for N fertilization by up to 27 kg N ha^{-1} yr^{-1} (Hansen et al. 2000b).

Cover crops have been implemented in Danish agriculture with a compulsory use on 6 percent of the area grown with winter and spring cereals, field peas, and rape. Among the allowed cover-crop species are grasses, crucifers, and chicory. The application of mineral fertilizer in the following year is reduced by 12 kg N ha^{-1} to account for increased soil N mineralization.

Crop Rotation Effects

Efficient utilization of crop rotation effects is particularly important in low-input farming systems, such as organic arable farming, where there is a much larger dependency on N$_2$-fixing crops, including pulses, green-manure crops, and cover crops.

Olesen et al. (2002) showed that rotations without a whole-year green manure crop produced the greatest total yield. Dry matter yields and N uptake in grains in this rotation were about 10 percent higher than in the rotation with a grass–clover ley in one year of four. Therefore, the yield benefits from the grass–clover ley could not compensate for the yield reduction as a result of leaving 25 percent of the rotation out of production.

The yield response from applying manure was the same with and without a green-manure crop. The RE_N of harvested grain for ammonium–N applied in manure was about 37 percent for cereal crops. A similar RE_N of 41 percent was found on average for an application of 50 kg of fertilizer N to winter wheat in on-farm fertilizer experiments carried out during 1991 to 1998 (Knudsen et al. 1999).

Winter wheat has a relatively high N demand, which in low-input farming systems can be partly supplied by a green-manure crop before the winter wheat. In the crop-rotation experiment, winter wheat and winter rye were grown just after the grass–clover green-manure crop. This gave quite different results at the different sites and years. A regression of wheat grain N uptake on accumulated N in the aboveground biomass when the green-manure crop was cut gave slopes of 0.04, 0.07, and 0.10 for the coarse sand, loamy sand, and sandy loam soils, respectively. The number of cuts varied from two to four, and if it is assumed that the N accumulation is less than half the aboveground N at the time of cutting as a result of internal recycling, then the recovery of the N in the green manure varies between 10 and 25 percent in the first year after the green-manure crop. The variation is probably linked with N leaching during the winter after sowing the winter wheat because the highest N leaching was found for the coarse sand (120 kg N ha^{-1} yr^{-1}) and the lowest for the sandy loam (35 kg N ha^{-1} yr^{-1}).

The aboveground N in the cover crop and weeds was measured by sampling in early November in the year before the spring barley. A multiple regression analysis was performed for spring barley grain N uptake against N in the biomass in November with manure application and previous crop as additional class variables. The cover crop's RE_N was taken as the slope of this regression. The results indicated greater RE_N for the coarse sand (72 percent) compared with the other soil types (49–49 percent) in rotations 1 and 2, where non N-fixing cover crops were used (ryegrass and chicory). The N recovery efficiency was greatest in the rotation, where a red and white clover was used in combination with ryegrass as a cover crop (70–77 percent). The use of aboveground biomass as an indicator of cover crop N probably overestimates the N utilization because there is often a substantial below-ground component in cover crops (Hansen et al. 2000c).

Conclusions

Over 15 years, the N surplus in Danish agriculture has been reduced by 235 Gg N yr^{-1} without reducing productivity. The result has been an increase in the overall N recovery efficiency from 42 to 52 percent, achieved primarily through a higher utilization of N in animal manures and a better consideration of pre-crop effects on yield potential and N supply. The lower N rates have led to a slight reduction in cereal grain protein content, but because the grain is used primarily for animal feed, this does not constitute a general problem.

The higher utilization of N in animal manure has been achieved by reducing losses during application and by better accounting for efficiency in the application year and in the following years. There is probably little scope for further increasing the NUE of animal manures, although a modest improvement may be obtained through manure digestion, slurry acidification, and improving technologies for slurry injection.

The pre-crop effects of legumes (grass–clover and pulse crops) and cover crops are now included in the Danish fertilizer recommendations. These recommendations are, however, based on average growing conditions of the legumes, and in practice the N fixation and thus the residual N effects vary considerably. There is a need to account for this more clearly in the estimates of crop N demand and also to account for soil and climate differences in the pre-crop effects.

Further improvement of N utilization needs to focus on a better determination of crop N demand at the farm, field, and subfield levels. Individual technologies for properly determining this N demand have failed to provide good estimates. A combination of technologies for measuring soil and crop characteristics with modeling of long-term crop rotation effects on N mineralization may increase the precision in determining optimal N rate and thus further increase NUE.

Literature Cited

Berntsen, J., A. Thomsen, K. Schelde, and O. M. Hansen. 2002. Ny strategi for GPS-gødskning med sensor. *Agrologisk* 3-02:26–27.

Broge, N. H., A. Thomsen, and P. B. Andersen. 2003. Stability of selected vegetation indices commonly employed as indicators of crop development— The red edge inflection point (REIP) and the ratio vegetation index (RVI). Pp. 146–149 in *Proceedings of the seminar on implementation of precision farming in practical agriculture.* DIAS report 100. Foulum, Denmark: Danish Institute of Agricultural Sciences.

Eriksen, J. 2001. Nitrate leaching and growth of cereal crops following cultivation of contrasting temporary grasslands. *Journal of Agricultural Science, Cambridge* 136:271–281.

Francis, G. S. 1995. Management practices for minimising nitrate leaching after ploughing temporary leguminous pastures in Canterbury, New Zealand. *Journal of Contaminant Hydrology* 20:313–327.

Grant, R., G. Blicher-Mathiesen, V. Jørgensen, A. Kyllingsbæk, H. D. Poulsen, C. Børsting, J. O. Jørgensen, J. S. Schou, E. S. Kristensen, J. Waagepetersen, and H. E. Mikkelsen. 2000. *Vandmiljøplan II - midtvejsevaluering.* Silkeborg, Denmark: Miljø-og Energiministeriet, Danmarks Miljøundersøgelser.

Hansen, E. M., J. Djurhuus, and K. Kristensen. 2000a. Nitrate leaching as affected by introduction or discontinuation of cover crop use. *Journal of Environmental Quality* 29:1110–1116.

Hansen, E. M., K. Kristensen, and J. Djurhuus. 2000b. Yield parameters as affected by introduction or discontinuation of cover crop use. *Agronomy Journal* 92:909–914.

Hansen, E. M., A. Kyllingsbæk, I. K. Thomsen, J. Djurhuus, K. Thorup-Kristensen, and

V. Jørgensen. 2000c. *Efterafgrøder. Dyrkning, kvælstofoptagelse, kvælstofudvaskning og eftervirkning.* DJF Rapport No. 37. Foulum, Denmark: Danish Institute of Agricultural Sciences.

Jensen, B., P. Sørensen, I. K. Thomsen, E. S. Jensen, and B. T. Christensen. 1999. Availability of nitrogen in ^{15}N-labeled ruminant manure components to successively grown crops. *Soil Science Society of America Journal* 63:416–426.

Jensen, E. S. 1991. Nitrogen accumulation and residual effects of nitrogen cover crops. *Acta Agriculturae Scandinavica* 41:333–344.

Jensen, E. S. 1992. The release and fate of nitrogen from cover-crop materials decomposing under field conditions. *Journal of Soil Science* 43:335–345.

Kirchmann, H., and A. Lundvall. 1993. Relationship between N immobilization and volatile fatty acids in soil after application of pig and cattle slurry. *Biology and Fertility of Soils* 15:161–164.

Knudsen, L., T. Birkmose, R. Hørfarter, O. M. Hansen, H. S. Østergaard, and K. L. Hansen. 1999. Gødskning og kalkning. Pp. 182–238 in *Oversigt over landsforsøgene*, edited by C. Å. Pedersen. Skejby, Denmark: Landbrugets Rådgivningscenter.

Kyllingsbæk, A. 2000. *Kvælstofbalancer og kvælstofoverskud i dansk landbrug 1979–1999.* DJF Rapport Markbrug no. 36. Foulum, Denmark: Danish Institute of Agricultural Sciences.

Olesen, J. E., L. N. Jørgensen, J. Petersen, and J. V. Mortensen. 2003. Effects of rates and timing of nitrogen fertiliser on disease control by fungicides in winter wheat. 1. Crop yield and nitrogen uptake. *Journal of Agricultural Science, Cambridge* 140:1–13.

Olesen, J. E., I. A. Rasmussen, M. Askegaard, and K. Kristensen. 2002. Whole-rotation dry matter and nitrogen grain yields from the first course of an organic farming crop rotation experiment. *Journal of Agricultural Science, Cambridge* 139:361–370.

Petersen, N. 2002. What is the reason for and the consequences of the variation in N-optimum? Pp. 57–62 in *Optimal nitrogen fertilization— tools for recommendation. Proceedings from NJF seminar 322,* edited by H. S. Østergaard, G. Fystro. and I. K. Thomsen. DIAS report, Plant Production No. 84. Tjele, Denmark: Danish Institute of Agricultural Sciences.

Sommer, S. G., E. Friis, A. Bach, and J. K. Schjørring. 1997. Ammonia volatilisation from pig slurry applied with trail hoses or broadspread to winter wheat: Effects of crop developmental stage, microclimate, and leaf ammonia absorption. *Journal of Environmental Quality* 26:1153–1160.

Sørensen, P., and M. Amato. 2002. Remineralisation and residual effects of N after application of pig slurry to soil. *European Journal of Agronomy* 16:81–95.

Sørensen, P., and E. S. Jensen. 1998. The use of ^{15}N labelling to study the turnover and utilization of ruminant manure N. *Biology and Fertility of Soils* 28:56–63.

Sørensen, P., E. S. Jensen, and N. E. Nielsen. 1994. The fate of ^{15}N-labelled organic nitrogen in sheep manure applied to soils of different texture under field conditions. *Plant and Soil* 162:39–47.

Sørensen, P., I. K. Thomsen, B. Jensen, and B. T. Christensen. 2002. Residual nitrogen effects of animal manure measured by ^{15}N, in *Optimal nitrogen fertilization— tools for recommendation. Proceedings from NJF seminar 322,* edited by H. S. Østergaard, G. Fystro, and I. K. Thomsen. DIAS report, Plant Production No. 84. Tjele, Denmark: Danish Institute of Agricultural Sciences.

Sørensen, P., F. P. Vinther, S. O. Petersen, J. Petersen, and I. Lund. 2003. Høj udnyttelse af gyllens kvælstof ved direkte nedfældning. *Grøn Viden, Markbrug* 281, 1–6. Tjele, Denmark: Danish Institute of Agricultural Sciences.

Thompson, R. B., J. C. Ryden, and D. R. Lockyer. 1987. Fate of nitrogen in cattle slurry following surface application or injection to grassland. *Journal of Soil Science* 38:689–700.

Thomsen, I. K. 1995. Cover crop and animal slurry in spring barley grown with straw incorporation. *Acta Agriculturae Scandinavica* 45:166–170.

Thomsen, I. K. 2001. Recovery of nitrogen from composted and anaerobically stored manure labelled with ^{15}N. *European Journal of Agronomy* 15:31–41.

Thomsen, I. K., and Jensen, E. S. 1994. Recovery of nitrogen by spring barley following incorporation of ^{15}N-labelled straw and cover crop material. *Agriculture, Ecosystems and Environment* 49:115–122.

Thomsen, I. K., V. Kjellerup, and B. Jensen. 1997. Crop uptake and leaching of ^{15}N applied in ruminant slurry with selectively labelled faeces and urine fractions. *Plant and Soil* 197:233–239.

Thomsen, I. K., P. Sørensen, J. Djurhuus, B. Stenberg, H. S. Østergaard, and B. T. Christensen. 2003. Bestemmelse af plantetilgængeligt kvælstof i jord tilført afgrøderester og husdyrgødning. DIAS report, Plant Production. Tjele, Denmark: Danish Institute of Agricultural Sciences.

Thorup-Kristensen, K. 1993. The effect of nitrogen cover crops on the nitrogen nutrition of a succeeding crop. I. Effects through mineralization and pre-emptive competition. *Acta Agriculturae Scandinavica* 43: 74–81.

PART IV
High-input Systems

10

Rice Systems in China with High Nitrogen Inputs

Roland Buresh, Shaobing Peng, Jianliang Huang,
Jianchang Yang, Guanghuo Wang, Xuhua Zhong,
and Yingbin Zou

Nitrogen fertilizer is a vital input for ensuring sufficient production of rice, the dominant staple food in Asia. An increase in the use of N fertilizer has paralleled a continuing increase in rice production to meet the demand of a growing population. China is the world's main rice-producing country, accounting for about 31 percent of global rice production in 2002 (FAO 2003). The high production of rice in China is achieved through high yields in irrigated ecosystems having an adequate supply of water and high rates of N fertilizer. About 20 percent of the global production of N fertilizer is used for rice in Asia.

About 93 percent of the rice-producing area in China is irrigated, and the average national yield of rough rice from 1998 to 2002 of 6.3 t ha^{-1} is among the highest in the world. Based on data from 1995 to 1997, 5.7 million tons of fertilizer N was consumed on rice in China (Dat Tran, FAO, personal communication, 2001). This corresponded to 7 percent of the total global consumption of fertilizer N and 37 percent of the global use of N fertilizer for rice production. The average rate of N application for rice production in China was 180 kg ha^{-1}, which is markedly higher than the world average and among the highest average national N rates for rice in the world. Rice production accounted for 24 percent of total N fertilizer use in China.

It has been established that the nitrogen use efficiency (NUE) is relatively low in irrigated lowland rice ecosystems. Losses of applied N fertilizer, particularly as gases, are typically higher in lowland rice ecosystems with saturated or flooded soil than in cropping systems with aerated soil (Zhu 1997). According to Li (1997), the apparent recovery efficiency (RE$_N$) of fertilizer N for rice in China was of the order of 0.30 to 0.35 kg of N taken up kg^{-1} N applied. Li (2000) observed, however, that the average RE$_N$ of rice in Jiangsu Province was only 0.20 kg kg^{-1}, significantly below the national average. This low RE$_N$ was due largely to the high N rates used by farmers in the area.

143

Available evidence indicates that NUE of rice production in China is very low, if not the lowest among the major rice-growing countries. Low NUE with high N loss has environmental consequences. Surface runoff of N can cause the eutrophication of lakes and rivers, and nitrate leaching can result in groundwater pollution. The area of surface water with eutrophication is increasing in China, partly as a result of poor NUE in crop production (Li 2000).

Approaches to Increase Nitrogen Use Efficiency

Tools for Managing Fertilizer Nitrogen

Because leaf N content is closely related to photosynthetic rate and biomass production, it is a sensitive indicator of changes in crop N demand within a growing season. A key for developing improved N management is therefore to establish a method for the rapid diagnosis of leaf N status. The chlorophyll meter (e.g., Minolta SPAD) provides one such simple, rapid, and nondestructive method for estimating leaf N content (Balasubramanian et al. 1999).

The relatively high price of the SPAD meter limits its use by individual farmers. A simple alternative method is a leaf color chart (LCC), which compares the light reflected from leaves and provides a measure of the associated leaf N. Several types of LCCs have been developed, including ones by Zhejiang Agricultural University, China; the University of California, Davis, California; and the International Rice Research Institute from a Japanese prototype. The range of green colors differs visually among these three LCCs. Yang et al. (2003), however, reported strong correlations among the scores of these three types of LCCs. The LCC can be calibrated with a SPAD to determine the critical color for specific rice cultivars under local growing conditions. The SPAD and LCC have been used to determine the timing of topdressing and to adjust the doses of N at preset times of N application.

Site-specific Nutrient Management

A site-specific nutrient management (SSNM) approach to management of fertilizer N for rice was developed in the mid-1990s and evaluated from 1997 to 2000 in 205 irrigated rice farms at eight sites in Asia, including one site in China (Dobermann et al. 2002). The approach aimed at dynamic field-specific management of N, P, and K fertilizer to optimize the supply and demand of nutrients. The need for N fertilizer was determined from the gap between the supply of N from indigenous sources, as measured with an N omission plot, and the demand of the rice crop for N, as estimated from the total N required by the crop to achieve a yield target for average climatic conditions. A decision support system provided, before planting, a splitting pattern for the estimated total N fertilizer requirement (Witt and Dobermann 2004). The predetermined N doses in the splitting pattern were then dynamically adjusted upward or downward based on

Table 10.1. Effect of site-specific nutrient management on nitrogen fertilizer use, yield, nitrogen use efficiency, and gross returns above fertilizer cost for rice at Jinhua, Zhejiang, China for six seasons from 1998–2000[1]

Parameter	Levels[4]	Treatment[2] SSNM	FFP	Δ[3]
N fertilizer (kg ha⁻¹)	All	126	171	−45
	ER	126	165	−39
	LR	126	177	−50
Grain yield (t ha⁻¹)	All	6.4	6.0	0.4
	ER	5.9	5.5	0.4
	LR	7.0	6.6	0.4
Agronomic efficiency of N (kg grain kg⁻¹ N)	All	12.5	6.8	5.7
	ER	11.3	6.4	4.9
	LR	13.7	7.2	6.5
Recovery efficiency of N (kg N kg⁻¹ N)	All	0.31	0.19	0.12
	ER	0.29	0.19	0.10
	LR	0.33	0.19	0.14
Gross returns above fertilizer costs (U.S.$ ha⁻¹)	All	941	852	89
	ER	864	782	82
	LR	1043	944	98

[1] From Wang et al. (2004).
[2] FFP, farmers' fertilizer practice; SSNM, site-specific nutrient management.
[3] Δ = SSNM−FFP.
[4] All, all six crops grown from 1998–2000; ER, early rice; LR, late rice.

either chlorophyll meter or the LCC readings at the preset times of N application (Witt et al. 2004).

The performance of SSNM, as compared with the existing farmers' fertilizer practice (FFP), was evaluated for three years (1998–2000) at 21 rice farms in Jinhua, Zhejiang Province (Wang et al. 2001, 2004). Two rice crops (early rice from April to July and late rice from July to October) were grown in each year with irrigation. Farmers used on average about 170 kg N ha⁻¹ in each season. Most farmers applied nearly all the N fertilizer in two large doses during the first two weeks after planting and then applied little N fertilizer thereafter (Wang et al. 2001). With SSNM, the N application was reduced by about 45 kg N ha⁻¹, averaged over six seasons, to 126 kg N ha⁻¹ (Table 10.1). With SSNM, the pre-plant N application was smaller than that for the farmers' practice; the topdressed N application at 7 to 14 days after transplanting (DAT) was small (about 30 kg N ha⁻¹), and N was applied between 20 to 55 DAT. The average fertilizer P and K rates were relatively similar for SSNM and FFP.

When averaged for six seasons in three years, SSNM increased grain yield by 0.4 t ha^{-1}, agronomic efficiency (AE) from 6.8 to 12.5 kg kg^{-1}, RE$_N$ from 0.19 to 0.31 kg kg^{-1}, and gross returns above fertilizer costs by an average of $89 U.S. ha^{-1} (Table 10.1). The NUE obtained with SSNM, however, remained below the AE of 20 kg kg^{-1} and RE$_N$ of 0.50 kg kg^{-1} achievable for irrigated rice with good crop management (Peng and Cassman 1998).

Real-time Nitrogen Management

The SSNM management of N fertilizer as developed and evaluated from 1998 to 2000 used a "fixed time-adjustable dose" approach. The time for topdressing N fertilizer was preset at a critical growth stage, and the SPAD or LCC was used only to adjust N fertilizer doses upward or downward at these preset times of N application. An alternative is the "real-time N management" approach, in which SPAD or LCC measurements were taken at 7- to 10-day intervals from 15 to 20 DAT to flowering on the most recent fully expanded leaves. A top dressing of about 20 to 45 kg N ha^{-1} is then applied whenever the SPAD or LCC value falls below a critical threshold (Peng et al. 1996). Real-time N management with the SPAD or LCC has now been evaluated in numerous farmers' fields in Asia since 1998, and the results often show that 20 to 30 percent less fertilizer N is required to achieve the same rice yield as obtained with the FFP (Bijay Singh et al. 2002).

Further Evaluation of Approaches to Increase Nitrogen Use Efficiency

Methods

Based on the encouraging results with SSNM as a promising technology for rice farmers (Table 10.1), the evaluation was expanded to three additional sites in China in 2001. The sites were in the major rice-growing provinces of Jiangsu, Hunan, and Guangdong. Rice production in Jiangsu typically involves one crop of japonica rice per year, which based on provincial statistics for 1998 through 2000 attains high average yield of 8.5 t ha^{-1} with high use of N fertilizer averaging 259-275 kg N ha^{-1}. Two crops of rice, often hybrids, are typically grown per year in Hunan and Guangdong.

Surveys of farmers' practices at the four study sites showed that the average rates of fertilizer N use ranged from 180 kg N ha^{-1} at the Hunan site to 240 kg N ha^{-1} at the Jiangsu site (Table 10.2). Surveys at the four sites indicated that the farmers apply 55 to 85 percent of the N as a basal dressing and a top dressing within the first 10 days after transplanting. The farmer's aim with the large applications of N during the early growing season is to reduce transplanting shock and stimulate early tillering.

In 2001 and 2002, the agronomic performance of N fertilizer strategies with different degrees of real-time N management was evaluated in a researcher-managed exper-

Table 10.2. Nitrogen rates (kg N ha^{-1}) and timing for each nitrogen fertilizer application in the farmers' fertilizer practice

Sites	1		2		3		4		5		Total N
	Rate	DAT[1]	Rate	DAT	Rate	DAT	Rate	DAT	Rate	DAT	
Jiangsu	144	0	24	4	36	46	36	66	—	—	240
Zhejiang	100	0	70	8	30	58	—	—	—	—	200
Hunan	100	0	35	19	25	35	20	64	—	—	180
Guangdong	70	0	65	5	30	24	20	37	15	59	200

[1] Timing of N application is indicated as days after transplanting (DAT). Basal N application is indicated as DAT = 0.

Table 10.3. Method for determining the rate of nitrogen application in "fixed time-adjustable dose" approach to site-specific nitrogen management at four sites in China

N application	Growth stage	% of total N	N rate (kg ha⁻¹)	If SPAD
1	Pre-plant	35	50	
2	Midtillering	20	30 ± 10	*
3	Panicle initiation	30	40 ± 10	**
4	Heading	15	± 20	***
Total		100	100–160	

*If SPAD > 36, apply 20 kg N ha⁻¹; if < 34, apply 40 kg N ha⁻¹; if between 34 and 36, apply 30 kg N ha⁻¹.
**If SPAD > 36, apply 30 kg N ha⁻¹; if < 34, apply 50 kg N ha⁻¹; if between 34 and 36, apply 40 kg N ha⁻¹.
***In favorable season and if SPAD < 36, apply 20 kg N ha⁻¹.

iment with four replications at each of the four provincial sites. Experiments were conducted in farmers' fields using an indica/indica hybrid cultivar, Shanyou 63, at all sites. Thirty-day-old seedlings were transplanted at 20×20-cm spacing with one seedling per hill. Phosphorus, potassium, and zinc were applied one day before transplanting to eliminate them as constraints to yield. An N omission plot (no added N fertilizer treatment) was included to estimate the indigenous N supply as determined from the total accumulation of N by rice when N fertilizer was not applied.

The four investigated N fertilizer strategies included the following:

1. FFP, based on the common practice near the sites (Table 10.2).
2. Modified FFP, derived by reducing total N input in FFP by 30 percent and restricting this reduction to within the first 10 DAT. (The modified FFP was specifically included to assess whether NUE could be increased by reducing the large early N applications used by farmers. It was hypothesized that reducing farmers' total N rate by 30 percent during early vegetative stage would not decrease the yield).
3. Real-time N management with the SPAD. SPAD measurements started at 10 DAT and continued at weekly intervals until heading. No basal N fertilizer was applied, and N fertilizer was applied at 30 kg N ha⁻¹ whenever the SPAD reading fell below 35 before panicle initiation. At the panicle initiation stage, N was applied once at 45 kg N ha⁻¹ when the SPAD fell below 35.
4. The fixed time-adjustable dose approach to SSNM (Table 10.3) comparable with that previously evaluated in Zhejiang (Table 10.1). The total N rate was preset based on the gap between the yield target and grain yield in the zero-N control. The timing of N application was fixed, but the doses for in-season N applications were adjusted upward or downward depending on leaf N status as determined with the SPAD (Table 10.3) (Witt et al. 2004).

Results

When averaged over the two years of this study, the indigenous N supply (INS) as determined from the total N accumulation by rice in N-omission plots ranged from 88 to 103 kg N ha^{-1}, which corresponded to grain yields of about 6 to 7 t ha^{-1}. Dobermann et al. (2003), in a comparison of INS in seven major irrigated rice-growing domains in Asia, found higher INS at the one China site in Jinhua, Zhejiang Province, than in the other six sites outside China. The measured INS averaged for 21 farm fields for eight rice crops was 69 kg N ha^{-1}, which corresponded to an average grain yield of 5 t ha^{-1}. The higher INS in our study compared with that reported by Dobermann et al. (2003) might in part be attributed to the growth of only one rather than two rice crops per year.

Among the four sites, grain yield was highest at Jiangsu. Yield averaged 8.8 t ha^{-1} for the two years with the FFP and increased to 9.7 t ha^{-1} with the modified FFP (Table 10.4). The AE of N fertilizer with the FFP was very low (< 7 kg kg^{-1}) at all sites, and AE was lower for the FFP than the other N treatments. The low AE with the FFP was attributed to the low response to N fertilizer—as a result of high INS—and the high rates of N fertilizer application.

The modified FFP successfully reduced N input by 30 percent without a reduction in yield (Table 10.4). In fact, the modified FFP increased yields by 10 percent at Jiangsu, 5 percent at Zhejiang, and 6 percent at Hunan, compared with the FFP. At Guangdong, the modified FFP had no effect on yield compared with the FFP. The AE was increased by the modified FFP at all sites except for Guangdong, where AE was low because of the low response to applied N fertilizer.

Real-time N management with the SPAD successfully reduced N inputs and increased yields by 5 percent at Jiangsu and Zhejiang and by 8 percent at Hunan compared with the FFP (Table 10.4). Real-time management markedly increased AE compared with both the FFP and the modified FFP. The high AE with real-time management was attributed to a large reduction in the rate of N fertilizer use. The total N rate for real-time management was only 38 to 90 kg ha^{-1}. Fertilizer N use was 97 to 150 kg N ha^{-1} less with real-time management than with the FFP. This corresponds to an N rate with real-time management of only 30 to 46 percent of the rate used with the FFP.

The fixed time-adjustable dose approach to SSNM also successfully reduced N inputs and increased yields by 9 percent at Jiangsu and Hunan and by 7 percent at Zhejiang compared with the FFP (Table 10.4). The AE markedly increased compared with both the FFP and the modified FFP. The fixed time-adjustable dose approach consistently matched or slightly exceeded real-time management in terms of yield. It generally matched the AE of real-time management except at Zhejiang, where yields were comparable but the N rate was higher with the fixed time-adjustable dose approach.

These findings show that considerable opportunity exists to increase NUE with

Table 10.4. Effect of nitrogen management practices on nitrogen fertilizer use, yield, and agronomic efficiency of nitrogen fertilizer (AE) at four locations in China averaged for two years (2001–2002)

	Farmers' fertilizer practice			Modified farmers' fertilizer practice			Real-time N management			Fixed time-adjustable dose N management		
	Fertilizer N (kg ha⁻¹)	Yield (t ha⁻¹)	AE (kg kg⁻¹)	Fertilizer N (kg ha⁻¹)	Yield (t ha⁻¹)	AEN (kg kg⁻¹)	Fertilizer N (kg ha⁻¹)	Yield (t ha⁻¹)	AE (kg kg⁻¹)	Fertilizer N (kg ha⁻¹)	Yield (t ha⁻¹)	AE (kg kg⁻¹)
Jiangsu	240	8.8	6.6	170	9.7	14.5	90	9.2	22.0	110	9.6	21.7
Zhejiang	200	6.5	1.4	140	6.8	4.4	60	6.8	13.2	110	6.9	6.5
Hunan	180	6.8	3.9	130	7.2	8.5	83	7.4	15.3	105	7.4	12.4
Guangdong	200	6.6	2.4	140	6.6	3.3	38	6.3	7.5	100	6.7	6.5

Unpublished data.

either no loss or a small gain in yield through site-specific N management with either a real-time or a fixed time-adjustable dose approach. The overapplication of N by farmers occurs mainly during the early vegetative stage. The large early applications of N fertilizer combined with the high INS and low demand of the young rice plant for N creates a situation ideal for large losses of N fertilizer from the rice fields to the external environment. The large early applications of N fertilizer and the high INS can also lead to luxury consumption of N by the rice plant, which can result in high-maintenance respiration, greater disease and pest damage, lodging, and low harvest index. Improved N management in which N inputs are better matched to crop needs minimizes the chance of luxury N consumption.

Conclusions

The results suggest that farmers in major rice-growing areas of China overapply N fertilizer. It is also evident that farmers do not consider the high indigenous supply of N in the field when they determine the level of N to apply to their rice crop. Our data from four sites suggest the agronomic efficiency of N fertilizer in China could be improved from 5 to 10 kg rough rice per kilogram of N applied without reducing rice production. Researchers and policy makers must work together to reach this goal.

The major task of researchers at this stage is to expand the on-farm demonstration of improved N management technology to convince farmers, researchers, and governmental officials that overapplication of N fertilizer is common and is a serious problem associated with rice production in China. Researchers should confirm through the conduct of medium-term experiments in farmers' fields that the reduction of N application through optimized N management will not reduce soil fertility and rice yield. Refinement and simplification should be done on the N management technology to facilitate adoption. The social and economic benefits of improved N management should be fully demonstrated. More important, government agricultural extension services and policy makers at all administrative levels must implement necessary policy interventions to accelerate the adoption of improved N management technology.

Acknowledgments

The unpublished data presented in this paper were obtained through the Reaching Toward Optimal Productivity (RTOP) work group of the Irrigated Rice Research Consortium (IRRC). Funding for RTOP was provided by the Swiss Agency for Development Cooperation (SDC), the International Fertilizer Industry Association (IFA), the Potash and Phosphate Institute/Potash and Phosphate Institute Canada (PPI/PPIC), and the International Potash Institute (IPI).

Literature Cited

Balasubramanian, V., A. C. Morales, R. T. Cruz, and S. Abdulrachman. 1999. On-farm adaptation of knowledge-intensive nitrogen management technologies for rice system. *Nutrient Cycling in Agroecosystems* 53:59–69.

Bijay Singh, Yadvinder Singh, J. K. Ladha, K. F. Bronson, V. Balasubramanian, J. Singh, and C. S. Khind. 2002. Chlorophyll meter- and leaf color chart-based nitrogen management for rice and wheat in northwestern India. *Agronomy Journal* 94:821–829.

Dobermann A., C. Witt, D. Dawe, S. Abdulrachman, H. C. Gines, R. Nagarajan, S. Satawathananont, T. T. Son, P. S. Tan, G. H. Wang, N. V. Chien, V. T. K. Thoa, C. V. Phung, P. Stalin, P. Muthukrishnan, V. Ravi, M. Babu, S. Chatuporn, J. Sookthongsa, Q. Sun, R. Fu, G. C. Simbahan, and M. A. A. Adviento. 2002. Site-specific nutrient management for intensive rice cropping systems in Asia. *Field Crops Research* 74:37–66.

Dobermann, A., C. Witt, S. Abdulrachman, H. C. Gines, R. Nagarajan, T. T. Son, P. S. Tan, G. H. Wang, N. V. Chien, V. T. K. Thoa, C. V. Phuong, P. Stalin, P. Muthukrishnan, V. Ravi, M. Babu, G. C. Simbahan, and M. A. A. Adviento. 2003. Soil fertility and indigenous nutrient supply in irrigated rice domains of Asia. *Agronomy Journal* 95:913–923.

FAO. 2003. Statistical databases, Food and Agriculture Organization (FAO) of the United Nations http://www.fao.org

Li, Q. 1997. *Fertilizer issues in the sustainable development of China agriculture.* Jiangxi Science and Technology Press (in Chinese).

Li, R. 2000. Efficiency and regulation of fertilizer nitrogen in high-yield farmland: A case study on rice and wheat double maturing system agriculture area of Tai lake for deducing to Jiangsu Province. Ph.D. Dissertation, China Agricultural University, Beijing, China (in Chinese).

Peng, S., F. V. Garcia, R. C. Laza, A. L. Sanico, R. M. Visperas, and K. G. Cassman. 1996. Increased N-use efficiency using a chlorophyll meter on high yielding irrigated rice. *Field Crops Research* 47:243–252.

Peng, S., and K. G. Cassman. 1998. Upper thresholds of nitrogen uptake rates and associated nitrogen fertilizer efficiencies in irrigated rice. *Agronomy Journal* 90:178–185.

Wang, G. H., A. Dobermann, C. Witt, Q. Z. Sun, and R. X. Fu. 2001. Performance of site-specific nutrient management for irrigated rice in southeast China. *Agronomy Journal* 93:869–878.

Wang G.H., Q. Sun, R. Fu, X. Huang, X. Ding, J. Wu, C. Huang, and A. Dobermann. 2004. Site-specific nutrient management in irrigated rice systems of Zhejiang Province, China. Pp. 243–263 in *Increasing productivity of intensive rice systems through site-specific nutrient management,* edited by A. Dobermann, C. Witt, and D. Dawe. Enfeld, NH: Science Publishers, and Los Baños, Philippines: International Rice Research Institute.

Witt, C., and A. Dobermann. 2004. Towards a decision support system for site-specific nutrient management. Pp. 359–395 in *Increasing productivity of intensive rice systems through site-specific nutrient management,* edited by A. Dobermann, C. Witt, and D. Dawe. Enfeld, NH: Science Publishers, and Los Baños, Philippines: International Rice Research Institute.

Witt, C, R. J. Buresh, V. Balasubramanian, D. Dawe, and A. Dobermann. 2004. Principles and promotion of site-specific nutrient management. Pp. 397–410 in *Increasing productivity of intensive rice systems through site-specific nutrient management,* edited by

A. Dobermann, C. Witt, and D. Dawe. Enfeld, NH: Science Publishers, and Los Baños, Philippines: International Rice Research Institute.

Yang W. H., S. Peng, J. Huang, A. L. Sanico, R. J. Buresh, and C. Witt. 2003. Using leaf color charts to estimate leaf nitrogen status of rice. *Agronomy Journal* 95:212–217.

Zhu, Z. 1997. Fate and management of fertilizer nitrogen in agro-ecosystems. Pp. 239–279 in *Nitrogen in soils of China*, edited by Z. Zhu, Q. Wen, and J. R. Freney. Dordrecht, The Netherlands: Kluwer Academic Publishers.

11

Using Advanced Technologies to Refine Nitrogen Management at the Farm Scale: A Case Study from the U.S. Midwest

T. Scott Murrell

Case Study Setting

The case study comes from Indiana, located in the east–central area of the U.S. Corn Belt. The Corn Belt is characterized by mean annual rainfall of 760 mm to more than 1000 mm, making irrigation unnecessary for crop production on most of the land area (Neld and Newmann 1990). The predominant rotation is maize/soybean (*Zea mays* L. / *Glycine max* (L.) Merr.), which is used on nearly all the arable land in this area (Christensen 2002).

Considering the United States as a whole, farms with the smallest maize acreage (100 ha or less) make up about 75 percent of all U.S. maize farms but produce only 29 percent of total national maize production (Foreman 2001). Nearly half of these small farms are located in the Heartland, defined by the Economic Research Service (ERS) as the eastern parts of South Dakota and Nebraska; southern and western Minnesota; northern and central Missouri; all of Iowa, Illinois, and Indiana; and the western half of Ohio (ERS 2000). Conversely, farms with more than 300 ha of planted maize constitute less than 4 percent of the maize farms in the United States but account for just under 20 percent of U.S. maize production (Foreman 2001).

A primary economic goal of farmers in the Corn Belt is to produce maize and soybean at the lowest cost per unit of production (kilograms of grain). Growing higher-yielding crops is important for reaching this goal, especially because many of the production costs cannot be cut back. Larger maize farms are characterized by higher expected yields, higher actual yields, and lower per-unit production costs (Foreman 2001).

According to a recent survey (Christensen 2002; ERS 2003), nearly every hectare of land planted to maize in the Corn Belt is fertilized with nitrogen (N). Approximately

19 percent of the area planted to maize received all of the N in the fall preceding the subsequent maize crop, whereas 32 percent of the area received N either in the spring before planting or split between the fall and spring. About 24 percent of the area received all the N at or after planting. The remaining 24 percent represented spring applications where N rates were split between planting and after-planting timing. Animal manure is applied to about 14 percent of the area planted to maize in this region.

The U.S. Midwest (U.S. Census Bureau 2003) is a U.S. leader in the adoption of many precision agriculture technologies. These include geo-referencing soil samples with global positioning system (GPS) receivers, using geographic information system (GIS) software to create sub-field scale agronomic recommendations, and using remote sensing for site characterization and monitoring crop progress (Whipker and Aldridge 2003).

In recent years, manure management has been receiving increasing attention throughout the United States. In 1998 through 1999, the U.S. Environmental Protection Agency (USEPA) and the U.S. Department of Agriculture (USDA) jointly developed and published the "Unified National Strategy for Animal Feeding Operations" (USDA and USEPA 1999). This strategy established national goals and performance expectations for all animal feeding operations (AFOs). A significant outcome of this strategy was the expectation that all AFOs develop and implement a comprehensive nutrient management plan (CNMP). The standard that pertains to nutrient management is Code 590. The unified strategy created stricter state-specific manure application guidelines that must be met in order for animal producers to receive a National Pollutant Discharge Elimination System (NPDES) permit (or state equivalent). These permits are required for the operation of certain types of AFOs.

It is in this setting of intensive maize and soybean production and increasing regulatory pressure that the case study is based. Discussion centers on approaches taken by an agronomist at a retail fertilizer outlet to tailor N management to local conditions. To examine the effectiveness of these practices, changes in maize grain yield and N use efficiency (NUE, kg grain kg N^{-1}) are presented for several fields from a 1180 ha family farm serviced by the outlet. In addition to maize and soybeans, this farm also markets about 5200 hogs yr^{-1} from a 2600-head confinement operation.

Development of Local Fertilizer Nitrogen Recommendations

The N management practices developed by the agronomist were part of an overall approach that attempted to manage nutrients variably within fields. Before instigating local research efforts on N, the agronomist had already established a site-specific management program in which soil types were used as the basis for creating management zones within fields. These zones were sampled separately to assess the chemical characteristics of soil. Lime, potassium (K), and phosphorus (P) were variably applied to different zones within fields, based on soil test results; however, N was still managed on a whole-field basis, with a uniform rate of 235 kg ha^{-1} being typical for most farmers in the area. As the ability to apply N variably across the field became feasible with new

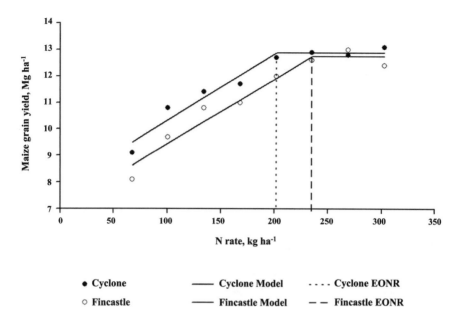

Figure 11.1. Maize yield response to incremental rates of N for the Fincastle and Cyclone soils with associated economically optimum N rates (EONR). Each data point represents the mean of 16 observations (four replications, four years).

technological developments, the agronomist began to research the economically optimum rates of N for the two dominant soil types in the area.

A study was established to investigate the response of maize to N on a Fincastle silt loam (fine-silty, mixed, mesic Aeric Ochraqualfs) and a Cyclone silt loam (fine-silty, mixed, mesic Typic Argiaquoll). A split-plot experiment, replicated four times, was designed with soil type as the whole plot and N rate as the subplot. N rates were selected to encompass local farmer management practices. The study was conducted for five years and was configured so that maize always followed soybean. This was done to make the results applicable to the maize/soybean rotations used in the area.

Four-year average responses (a drought year omitted) were analyzed and economically optimum N rates (EONR) determined using a linear-plateau model (Anderson and Nelson 1975) (Figure 11.1). The EONRs were 200 kg N ha^{-1} for the Cyclone soil and 235 kg N ha^{-1} for the Fincastle soil. These N recommendations represented a different approach to N fertilization than that recommended by the Cooperative Extension Service at Purdue University. The university recommendation system increased N rates according to yield goals (Vitosh et al. 1995). The local recommendations used only two rates, one appropriate for the average responses observed on each of the two dominant soil types. In most cases, based on yield goals in the agronomist's geographic area, the locally developed N recommendation rates were lower than those from Purdue. This may have been due in part to the choice of a response model that was possibly more conservative in its determination of EONRs (Cerrato and Blackmer 1990). The local recommendations were also counter to the opinion held by many farmers in the area that the darker Cyclone soil should get more N because it was more productive. Local research results indicated that the Cyclone soil actually required less N to achieve economically optimum yields. Results from this study were inferred to analogous soil types in the area: a Crosby silt loam (fine-loamy, mixed, mesic Aeric Ochraqualfs) and a Brookston silt loam (fine-loamy, mixed, noncalcarious, mesic Typic Argiudolls). With completion of this study, the agronomist began varying N rates across farmers' fields using GPS and single-product rate controllers.

At about the same time the N rate study was being conducted, the agronomist initiated another experiment that examined N timing. Before the agronomist's employment at the dealership, the retail outlet promoted spring applications of N. This was done not for agronomic reasons, but because spring was the time when the dealership's inventory of N was greatest. Most customers were applying N in the spring before maize was planted. Results from an eight-year study (not presented) showed a consistent (in seven of eight years) yield advantage to applying the same rate of N at a later date, closer to the time of crop need (termed *side-dressing*).

In the past, side-dressed N applications were much more widely used than they are now. This practice involved no additional monetary investment by the farmers and was quickly adopted. By the third year of the study, the agronomist estimates that about 60 percent of the trade area was receiving N at this time in the season. Although farmers were not taking on any additional financial risk with this practice, the dealership was because it was responsible for the timely application of the N. As the practice began to be accepted, the retail outlet found itself facing high demand for its custom N application business during a narrow window of time. The demand for side-dressed N outpaced the dealership's resources to provide the service. Eventually, many of the sales staff no longer promoted the practice, and adoption declined.

Open dialogue between the agronomist and farmer customers, combined with the previously established site-specific management program, was critical to implementing new N management practices. Variable N applications based on soil type are used on approximately 10,000 ha (about 25 percent of the dealership's trade area). Variable N applications combined with side-dressed N applications are used on approximately 5000 ha (13 percent of the dealership's trade area).

Effects of Local Fertilizer Nitrogen Management at the Farm Scale

The producers from the 1180-ha case study farm have traditionally worked closely with the agronomist and were some of the first adopters of his new recommendations. Previously, these farmers applied a uniform rate of 235 kg N ha^{-1} across every field. In Fig-

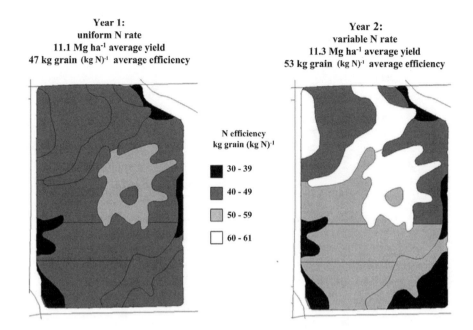

Year 1:
uniform N rate
11.1 Mg ha⁻¹ average yield
47 kg grain (kg N)⁻¹ average efficiency

Year 2:
variable N rate
11.3 Mg ha⁻¹ average yield
53 kg grain (kg N)⁻¹ average efficiency

N efficiency
kg grain (kg N)⁻¹

30 - 39
40 - 49
50 - 59
60 - 61

Figure 11.2. Comparison of N efficiency differences between two different but similarly yielding years on the same field. In year 1, a single rate of 235 kg N ha⁻¹ was applied uniformly across the field. In year 2, N rate was varied by soil type.

ure 11.2, year 1 is an example of how N efficiencies varied across the field when using a uniform N rate. NUEs ranged from 39 to 52 kg grain kg N⁻¹, with an area-weighted average N efficiency of 47 kg of grain kg N⁻¹. Year 2 shows how NUE improved in many areas of the field when N rates were varied by soil type. Nitrogen efficiencies ranged from 36 to 67 kg of grain kg N⁻¹ with an area-weighted average NUE of 53 kg of grain kg N⁻¹. Because field average yield levels were nearly the same in both years, the 6 kg grain N kg N⁻¹ average increase in efficiency resulted primarily from the reduced N rate of 200 kg ha⁻¹ applied to the Brookston soil. The Cyclone and Brookston soils receiving less N constitute about 53 percent of the total land area of the farm not receiving manure applications.

Higher efficiencies were also attained by increasing yields. To examine temporal trends in yields on the case study farm, 37 of 52 fields were selected, based on their longer history of management with variable N and the fact that they had not received manure. Each field was harvested using a yield monitor coupled to a differentially corrected GPS receiver. Using GIS software, yield monitor data were averaged over contiguous soil type regions in each field. On average, 381 ha from 17 maize fields were analyzed each year.

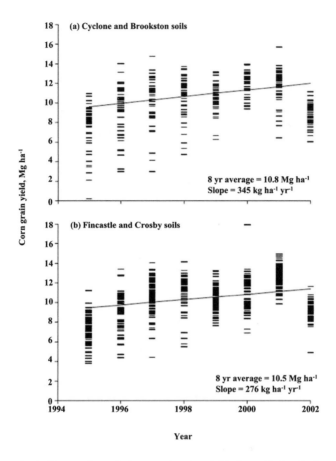

Figure 11.3. Temporal trends in annual maize yields from (a) the Cyclone and Brookston soils and (b) the Fincastle and Crosby soils.

Figure 11.3 shows that over the eight-year period considered, yields significantly increased (p value < 0.01) on both groups of soils. Two drought years occurred during this period, one in 1995 and another in 2002. The 345 kg ha^{-1} yr^{-1} average annual yield increase on the Cyclone and Brookston soils was slightly but significantly higher than the 276 kg ha^{-1} yr^{-1} yield increase on the Fincastle and Crosby soils (p value = 0.056). The average yield across all years for the Cyclone and Brookston soils was 10.8 Mg ha^{-1}, which was significantly higher than the 10.5 Mg ha^{-1} average yield of the Fincastle and Crosby soils (p value < 0.01).

Increasing yields resulted in significantly greater N efficiencies over time (Figure 11.4). Within each soil group, applied N rates remained constant over the period considered: 200 kg N ha^{-1} for the Cyclone and Brookston soils and 235 kg N ha^{-1} for the

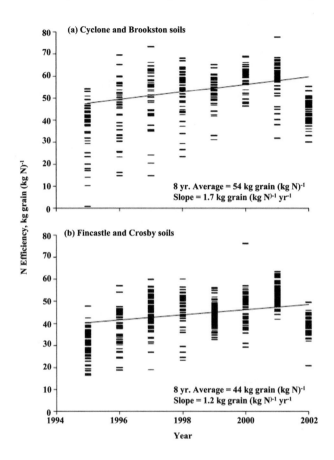

Figure 11.4. Temporal trends in annual N efficiency from (a) the Cyclone and Brookston soils and (b) the Fincastle and Crosby soils.

Fincastle and Crosby soils. The Cyclone soils exhibited N efficiencies that increased by about 1.7 kg grain kg N^{-1} yr^{-1}, which was slightly but significantly more rapid than the 1.2 kg grain kg N^{-1} yr^{-1} observed on the Fincastle and Crosby soils (p value < 0.01).

The differences in NUE between soil groups over time in Figure 11.4 demonstrate the net effect of increasing yields and simultaneously reducing N rates. Over the eight-year period, N efficiencies on the Cyclone and Brookston soils ranged from 0.85 to 78.0 kg of grain kg N^{-1} and averaged 54 kg of grain kg N^{-1}. NUE on the Fincastle and Crosby soils ranged from 16 to 76 kg of grain kg N^{-1} and averaged 44 kg of grain kg N^{-1} during the eight years. Whereas the average N efficiency on the Cyclone and Brookston soils was significantly higher (p value < 0.01), variability in NUE was also significantly higher on these soils (p value < 0.01).

Increasing yields and NUE have resulted not only from improved N management but also from the entire set of management practices used by the farmers. Specific to nutrient management, variable N applications are part of a broader site-specific management program that includes variable P, K, and lime applications. The nutrient management approach recommended by the university and adopted by agronomists is to build soil test P, K, and pH to levels that do not limit crop yields (Vitosh et al. 1995). Reduced tillage (Conservation Technology Information Center 2003) is used on all non-manured fields. Like many in the area, the farmers have adopted new glyphosate-resistant soybean varieties and new maize hybrids. Therefore, the newly adopted N management strategies are part of a much larger management system.

Development of Local Manure Nitrogen Management Practices

Interest in integrating commercial fertilizer with manure applications has stemmed largely from recent changes in governmental regulations. In Indiana, the Indiana Department of Environmental Management (IDEM) regulates confined feeding operations (USEPA 2002). The state requires that permits be renewed every five years. The family farmers in the case study currently have permits for their hog operation but must renew them again in the future under stricter requirements. For this reason, the agronomist and farmers have begun to communicate about how manure management practices need to be altered. In addition, the farmers are also interested in participating in Natural Resources Conservation Service programs that provide funding to farmers who want to implement improved conservation practices, including better management of manure nutrients.

One of the first activities by the farmers and agronomist to improve manure management was calibrating the manure applicator to determine application rate. The agronomist used software running on a hand-held computer coupled to a GPS receiver to record the distance traveled by the applicator. This distance was then converted to an area and an average application rate calculated. The manure applicator as operated applies approximately 69,000 L ha^{-1}. At this rate, about 36 percent of the total manure generated annually is applied at rates that supply N at 65 to 73 percent of that allowed under the new permit requirements. The remaining 64 percent of the total annual production of manure has higher N concentrations that require application rates as low as 36,500 L ha^{-1}, or approximately 53 percent of the manure rate currently applied by the farmers. This rate is beyond the capabilities of the family farmers' current equipment. The only solution to this problem is to purchase new equipment, which will be a significant investment.

At the same time manure application practices were being characterized, the agronomist and farmers were working together to model the hog operation. This was facilitated by prototype software being developed by Purdue University to assist with manure management planning (Joern and Hess 2003). Manure-application schedules that must be employed to keep pits from becoming too full require application times that are not always well timed with crop uptake, such as summer and fall. This means that an unknown portion of the N supplied by manure and required by maize the next year will be lost before the crop is grown.

To determine whether supplemental fertilizer N should be applied to manured fields in the spring when maize is grown, the agronomist is adopting the presidedress soil nitrate test (PSNT) procedures outlined by researchers at Purdue (Brouder and Mengel 2003). This test uses in-season assessments of soil nitrate levels to adjust recommended N rates for maize. The PSNT samples will be taken in the spring from each geo-referenced zone defined by the area spread with manure from each storage pit. This practice is being tried on all manured fields.

Summary and Comments

This case study is from the U.S. Midwest, an area characterized by intensive production of maize and soybean. Most maize production comes from a minority of larger farms. These farms target and reach higher yields than the smaller farms, allowing them to produce maize at lower per-unit costs. Nitrogen is applied to nearly every hectare planted to maize.

The case study focuses on an innovative agronomist working at a retail fertilizer outlet and one of his progressive farmer customers. Fertilizer N management practices were developed by the agronomist through a locally based research program. The objective of the agronomist was to determine how N management practices needed to be changed to better fit local conditions. Years of collecting data on local management practices, coupled with replicated research trials, allowed him to determine economically optimum N rates appropriate for the soils in his geographic area. The research also produced recommendations for N applied early in the season, during early maize growth stages. The farmers in the case study have adopted both of these improved N management practices.

The impacts of these locally developed practices have been positive. A nitrogen rate that is 86 percent of that used previously is being applied to just over half of the cropped area. The N management approaches are part of a larger site-specific nutrient management strategy that builds and maintains soil fertility at levels considered nonlimiting to crop production.

The following are the major technologies that have been used to improve yields and NUE:

• Computers with improved capabilities
• Geographic information system software
• Statistics and spreadsheet software
• Manure management software

- Global positioning system receivers
- Yield monitors
- Variable rate controllers
- Soil testing, including the PSNT
- Manure nutrient testing
- Calibrated manure application equipment
- New soybean varieties, specifically genetically modified organisms
- New maize hybrids
- Field equipment for high-residue management

Until recent regulatory pressure, poor communication existed between the farmers and agronomist regarding manure applications; however, now that the agronomist and farmers are working together on manure management issues, many improvements have occurred within a short period. Zones defined by the area spread with manure from each storage pit define where samples will be taken in the spring to assess soil nitrate levels and adjust fertilizer N rates accordingly. Prototype software is being used to record information relevant to manure management and to create improved strategies.

While manure management is undergoing many positive changes, progress will be limited until new manure application equipment can be purchased. The current applicator has a narrow range of application rates. The rate currently used applies N at acceptable rates for the lower-analysis manure constituting approximately 36 percent of the total manure generated annually. The remaining higher-analysis manure must, to be within compliance, be applied at about 53 percent of the rate currently used by the case-study farmers. This is well beyond the capabilities of their equipment.

An important theme throughout the case study has been finding local solutions to local problems. Localized N recommendations and current improvements in integrating manure and fertilizer N are all examples of first taking inventory of what management practices currently exist and then devising ways of improving them. New approaches have been adopted because farmers were involved throughout the discovery process, the agronomist was reputable, and the research was local. Regulations and the potential for incentive payments for improved practices contributed to positive change. For government programs to have such desirable effects, however, they must be flexible enough to allow local solutions to be discovered and implemented to address local problems and improve local management practices in affordable ways. Although the solutions developed in this case study are primarily of local interest, the principles involved and particularly the range of modern technologies employed have very much wider application.

Literature Cited

Anderson, R. L., and L. A. Nelson. 1975. A family of models involving intersecting straight lines and concomitant experimental designs useful in evaluating response to fertilizer nutrients. *Biometrics* 31:303–318.

Brouder, S. M., and D. B. Mengel. 2003. *The presidedress soil nitrate test for improving N management in corn.* AY-314-W. http://www.agcom.purdue.edu/AgCom/Pubs/agronomy.htm. Lafayette, IN: Purdue University Cooperative Extension Service.

Cerrato, M. E., and A. M. Blackmer. 1990. Comparison of models for describing corn yield response to nitrogen fertilizer. *Agronomy Journal* 82:138–143.

Christensen, L. A. 2002. *Soil, nutrient, and water management systems used in U.S. corn production.* Agriculture Information Bulletin No. 774. Washington, D.C.: United States Department of Agriculture, Economic Research Service.

Conservation Technology Information Center. 2003. *Tillage type definitions.* http://www.ctic.purdue.edu/Core4/CT/Definitions.html.

ERS (Economic Research Service). 2000. *Farm resource regions.* Agriculture Information Bulletin No. 760 (September 2002). http://www.ers.usda.gov/publications/aib760/. Washington, D.C.: United States Department of Agriculture Economic Research Service.

ERS (Economic Research Service). 2003. *Agricultural resource management survey (ARMS).* http://www.ers.usda.gov/Briefing/ARMS/.

Foreman, L. F. 2001. *Characteristics and production costs of U.S. corn farms.* Agriculture Information Bulletin No. 974 (August 2001). http://www.ers.usda.gov/publications/sb974-1/sb974-1.pdf. Washington, D.C.: United States Department of Agriculture Economic Research Service.

Joern, B., and P. Hess. 2003. *Manure management planner.* http://www.agry.purdue.edu/mmp/.

Neld, R. E., and J. E. Newman. 1990. *Growing season characteristics and requirements in the Corn Belt.* National Corn Handbook NCH-40, http://www.ces.purdue.edu/extmedia/NCH/NCH-40.html. Lafayette IN: Purdue University Extension Service.

U.S. Census Bureau. 2003. *Census regions and divisions of the United States.* http://www.census.gov/geo/www/maps/CP_MapProducts.htm.

USDA and USEPA (United States Department of Agriculture and United States Environmental Protection Agency). 1999. *Unified national strategy for animal feeding operations* (March 9, 1999). http://www.epa.gov/npdes/pubs/finafost.pdf.

USEPA (United States Environmental Protection Agency). 2002. *State compendium—region 5: Programs and regulatory activities related to animal feeding operations* (May 2002). http://cfpub.epa.gov/npdes/afo/statecompend.cfm.

Vitosh, M. L., J. W. Johnson, and D. B. Mengel. 1995. *Tri-state fertilizer recommendations for corn, soybeans, wheat, & alfalfa.* Extension Bulletin E-2567. http://www.ces.purdue.edu/extmedia/AY/AY-9-32.pdf. Michigan State University, The Ohio State University, and Purdue University.

Whipker, L. D., and J. T. Adridge. 2003. *Precision agricultural services dealership survey results.* Staff Paper No. 3-10 (June 2003). http://www2.agriculture.purdue.edu/ssmc/. Lafayette, IN: Purdue University, Department of Agricultural Economics.

12

Impact of Management Systems on Fertilizer Nitrogen Use Efficiency

John Havlin

Although world food production doubled over the last 30 years, nitrogen (N) use increased sevenfold (Tilman 1999). Total cereal production and average yield must increase by nearly 50 and 40 percent, respectively, to meet world food demand in 2025 (FAO 2001). As a result, world N use will continue to increase. Assuming only modest increases in agricultural land use, future food demand requires greater production per unit of land. Without significant advances in N use efficiency (NUE), the N required to increase yield by 40 percent may further degrade water and air quality.

Numerous factors influence crop N requirement. In general, low NUE occurs when applied N exceeds yield potential. Increasing NUE requires improved N management that reflects natural N transformations affecting N loss and accumulation. Whereas advances in crop genetics will increase yield potential, accurately quantifying N requirements is a challenge met only through advances in science and in relentless educational efforts to encourage adoption of technologies to increase yield and NUE.

Long-term N studies provide insights into barriers to increasing NUE. With increasing N rate, wheat and maize yield increased while NUE decreased (Figure 12.1). No correlation existed between unfertilized and fertilized yields, suggesting that environmental conditions conducive to increasing native available N and yield potential do not necessarily increase supplemental N requirement.

Conventional Nitrogen Recommendations

Crop N requirement (N_{REQ}) is determined by the following:

$$N_{REQ} = N_{CROP} - N_{SOIL} \qquad \text{[Eq. 1]}$$

where, $N_{CROP} \rightarrow$ yield goal (kg ha^{-1}) $\times$ N coefficient (kg N kg yield^{-1}) = kg N ha^{-1} and $N_{SOIL} = [N_{sources} - N_{losses}]$ defined below. N_{CROP} varies between crops, soils, and

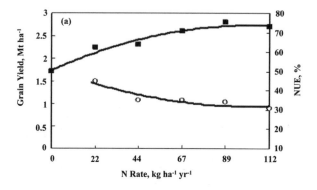

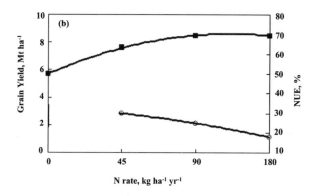

Figure 12.1. Effect of annually applied fertilizer N on average yield (■) and NUE (o) of winter wheat (1971–2000, Lahoma, Oklahoma) (a) and irrigated corn (1969–1983, Mead, Nebraska) (b). In the wheat study, 7-86 kg N ha^{-1} N was applied over that needed for optimum yield (Raun and Johnson 1995; Olsen et al. 1986).

regions, or climates. In corn, the N coefficient averages 0.013 kg N kg grain^{-1}, but it ranges between 0.016 and 0.03 kg N kg grain^{-1}. In wheat, 0.02 to 0.03 kg N kg^{-1} grain is typical. Thus, for 4 Mt ha^{-1} wheat yield goal is as follows:

$$N_{CROP} = 4000 \text{ kg ha}^{-1} \times 0.025 \text{ kg N kg}^{-1} = 100 \text{ kg N ha}^{-1}$$

Raun and Johnson (1995) also showed that the N coefficient varied from 0.007 to 0.056 kg N kg^{-1} grain; thus, using average expected yield and N coefficients will both underestimate and overestimate N_{REQ}. Overestimating yield goal will obviously overestimate N_{REQ} and reduce NUE. Surveys suggest that approximately 80 percent of producers overestimate yield goal (Schepers et al. 1986).

N_{SOIL} (Eq. 1) represents the soil's capacity to provide available N during the growing season and is defined by $N_{SOURCES} - N_{LOSSES}$. Equation 1 becomes:

$$N_{REQ} = N_{CROP} - [N_{SOURCES} - N_{LOSSES}] \qquad [Eq. 2]$$

where $N_{SOURCES} = N_{inorg} + N_{OM} + N_{sym} + N_{man} + N_{nonsym} + N_{depot}$

N_{inorg}	$\rightarrow$	residual inorganic N content in the soil profile
N_{OM}	$\rightarrow$	mineralizable N from soil organic matter (OM)
N_{sym}	$\rightarrow$	symbiotic N fixation
N_{man}	$\rightarrow$	N credit for previous waste application
N_{nonsym}	$\rightarrow$	non-symbiotic microbial N fixation
N_{depot}	$\rightarrow$	N deposition in rain or irrigation water

and $N_{LOSSES} = N_{leach} + N_{denitr} + N_{vol} + N_{eros} + N_{immob}$

N_{leach}	$\rightarrow$	N leached
N_{denitr}	$\rightarrow$	N denitrified
N_{vol}	$\rightarrow$	N volatilized
N_{eros}	$\rightarrow$	N loss through soil erosion
N_{immob}	$\rightarrow$	N immobilized (loss of plant available N)

Nitrogen Sources

N_{inorg} is commonly used where annual precipitation is less than 750 mm, whereas in regions with more than 750 mm, water transport below the root zone and denitrification of residual NO_3^- during non-crop periods reduces N_{inorg} to low levels. Thus, preplant N_{inorg} assessments by soil analysis are often unreliable in estimating N_{REQ}.

Few N_{REQ} models explicitly include N_{OM}, when a substantial portion of biomass N is due to N_{OM} (Rice and Havlin 1994). Omission of N_{OM} in N_{REQ} is due to our inability to provide preseason estimates of N_{OM}, although many soil and environmental factors influencing N_{OM} are understood. Vigil et al. (2002) showed N_{OM} ranges between 10 and 100 kg N ha^{-1} during the growing season.

Unfortunately, these estimates do not account for temporal and spatial variability in N_{OM}. The inability to account for temporal variability in N_{OM} and yield potential results in low NUE when average N_{REQ} is used.

The availability of N from previous legume crops (N_{sym}) depends on the quantity of N fixation and environmental conditions influencing N_{OM}. Most references report total legume N fixed, but few provide the proportion of N fixed left in the field after harvest, which is essential for estimating N_{REQ} of subsequent crops (Schepers and Mosier 1991). Pre-plant N_{inorg} does not account for N availability from previous legume crops. Whereas N_{sym} ranges between 20 and 350 kg N ha^{-1} for grain and forage legumes, residual N availability to subsequent crops is less than 150 to 200 kg N ha^{-1} (Giller et al. 1997). Estimates of N_{sym} availability are based on field trails measuring N uptake in unfertilized nonlegume crops following a legume crop. For corn following most forage legumes, N_{sym} ranges from 80 to 150 kg N ha^{-1} and 30 to 50 kg N ha^{-1} in the first and second years, respectively (Schepers and Fox 1989). N_{sym} for crops following soybean ranges between 20 and 50 kg N ha^{-1} (Kurtz et al. 1984). Estimates of N_{nonsym} vary widely. Under optimum conditions (high surface-soil moisture and C:N residue cover, low soil N), N_{nonsym} can reach 30 kg N $ha^{-1} yr^{-1}$. Under normal (periodic dry soil) conditions N_{nonsym} likely is less than 5 kg N $ha^{-1} yr^{-1}$.

Near industrial N emissions, total N_{depot} can reach 15 kg N $ha^{-1} yr^{-1}$, but it generally contributes 2 to 10 kg N $ha^{-1} yr^{-1}$, depending on the region. Irrigation water NO_3^- must also be credited, ranging from 10 to 145 kg N ha^{-1}, depending on N concentration and irrigation rate (Meisinger and Randall 1991).

Estimating residual N_{man} availability is more difficult than other mineralizable N sources because of spatial variability in waste application, variable N content between sources, variable N_{vol} losses during manure handling and application, and variable N_{denitr} losses (Schepers and Fox 1989). Estimates of first- and second-year N_{man} range between 20 and 90 percent and between 2 and 30 percent of total N applied, depending on manure type, rate, and application method. Most commercial livestock operations annually apply manure based on N_{REQ}. In these situations, additional fertilizer N is not required because of high N_{man}. Meisinger et al. (1992) reported N mineralization rates ranging between 0.1 and 1.5 mg N kg $soil^{-1} d^{-1}$, depending on manure N rate. Using an average 0.8 mg N $kg^{-1} d^{-1}$, about 80 kg N ha^{-1} would have been mineralized between corn planting and V6 growth stage (about 30 days). Decomposable manure C provides an energy source for denitrifiers, causing higher N_{denitr} in manured soils (Firestone 1982). Under optimum conditions for N_{denitr} (high soil moisture and surface residue cover), N losses can be 50 percent or less of applied manure N. Because of difficulties in establishing average N_{man} values, accurately estimating N_{REQ} in fields with past manure applications will be difficult, again because of uncertainty in predicting environment conditions controlling N mineralization.

Nitrogen Losses

N_{leach} is a major N loss pathway in aerated agricultural systems. Under tile drainage, N_{leach} can approach 60 percent of applied N, whereas under natural drainage values between 10 and 30 percent are common (Meisinger and Delgado 2002). In general, increased N_{leach} potential is related to N rates exceeding crop yield potential.

Timing N applications to avoid periods of high water transport reduces N_{leach} (Peoples et al., Chapter 4, this volume). Randall and Mulla (2002) reported an increase of greater than 20 percent in NUE with spring versus fall application of N. N_{inorg} can be reduced by 10 to 30 percent with legumes in the rotation (Kanwar et al. 1997) and 20 to 80 percent with cover crops (Dabney et al. 2001).

N_{denitr} varies depending on soil water content, inorganic soil N content, organic C supply, soil pH, and temperature (Peoples et al., Chapter 4, this volume). When soil water-filled pore space is 60 percent or greater, N_{denitr} increases. N_{denitr} estimates vary widely but are usually less than 15 percent of applied N. Peoples et al. (1995) reported losses of 1 kg N ha^{-1} d^{-1} under high N_{inorg}, temperature, and water content. Meisinger and Randall (1991) showed that N_{denitr} ranged from 2 to 25 percent of N applied in well-drained soils compared with 6 to 55 percent on poorly drained soils. N_{vol} occurs predominately with surface applied N to neutral and high pH soils and is markedly affected by environmental conditions (Peoples et al., Chapter 4, this volume). In flooded rice systems, N_{vol} can exceed 75 percent of applied N (Peoples et al. 1995). Typical N_{vol} in arable systems is usually less than 25 percent (Meisinger and Randall 1991).

N_{eros} contributes to the degradation of surface waters, depending on the quantity of soil loss and soil N content (Peoples et al., Chapter 4, this volume). Blevins et al. (1996) estimated that less than 15 percent of applied N is in runoff, which would vary greatly with N application method and the timing of runoff events.

N_{immob} does not represent a true N loss but rather a reduction in plant available as N_{inorg} is converted to organic N by microorganisms. N_{immob} increases with the increasing quantity of residue and decreasing residue N content. Microbes degrading residues containing less than 1.5 percent N ($\geq$ 30 C:N) generally immobilize inorganic soil N. Depending on residue quantity and N content, 20 to 50 percent of fertilizer N can be incorporated into soil OM (Power and Broadbent 1989). Fertilizer N placement below surface crop residues compared with broadcast N reduces N_{immob}, N_{vol}, and N_{denitr} while enhancing NUE.

From this discussion, the temporal and spatial variability in $N_{SOURCES}$ and N_{LOSSES} limits our ability to quantify N_{CROP}, N_{SOIL}, and ultimately N_{REQ} accurately. Efforts to increase NUE in cropping systems throughout the world will require improved tools to define yield potential and N availability from N_{OM}, N_{sym}, and N_{man} (Rice et al. 1995).

Nitrogen Recommendation Based on Average Yields

Most N_{REQ} systems are based on field trails that quantify crop response to applied N. N response data over many soils, soil and crop management inputs, and years are combined to develop N_{REQ} from average yield goals and N efficiencies (Eq. 1). Actual N rate needed for optimum yield varies greatly between years. Figure 12.2 shows typical crop N response variation (Bock and Hergert 1991). Based on annual optimum N rate, a

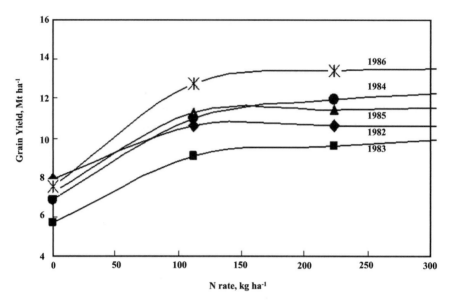

Figure 12.2. Variation in irrigated corn yield response to N in Nebraska (adapted from Bock and Hergert, 1991).

twofold range in yield per unit of N applied (N_{eff}) was observed (Table 12.1). At the optimum N rate, NUE ranged from 36 to 63 percent. Averaging these data gives 11.6 Mt ha^{-1} yield at 162 kg N ha^{-1}. To ensure that N is not limiting in high yield years, average yield potential is usually increased. Thus, increasing optimum yield to 13.0 Mt ha^{-1} resulted in an average 10 percent decrease in N_{eff} and NUE of in three of five years. This estimate is conservative because the N_{REQ} model would have resulted in 230 kg N ha^{-1}, reducing NUE from 44 to 30 percent.

Long-term N response data for wheat show that when N rates were based on average yield potential (2.7 Mt ha^{-1} yield $\times$ 0.033 kg N kg grain^{-1}), average yield goal was obtained 30 percent of the time (Johnson and Raun 2003). More importantly, 37 percent of the time an additional 23 kg N ha^{-1} was needed to optimize yield, and 33 percent of the time 23 to 90 kg ha^{-1} more N was applied than required for optimum yield. These data illustrate the difficulty in accurately estimating annual yield potential. Use of average yield estimates results in misapplication of N that reduces NUE. While year-to-year variation in environment contributes greatly to the error in predicting yield potential, these conditions also greatly influence N_{OM}. Many studies show decreased NUE with increasing contribution of N_{OM} to N supply and yield. Therefore, methods used to improve estimated annual yield potential by inclusion of soil or crop measures that capture between-year variability in N_{OM} could provide significant increases in NUE.

Table 12.1. Variation in irrigated corn yield response to nitrogen, nitrogen efficiency, and nitrogen use efficiency (NUE) between years. Grain N content of 1.2 and 1.4% was assumed for the unfertilized and fertilized treatments, respectively[1]

Year	Optimum Yield a	Yield No N b	Optimum N rate c	N_{eff} (optimum)[2] a/c	N_{eff} (Ave + 10%)[3] $a/c_{(ave+10\%)}$	NUE (optimum)[4] $(a_{1.4}-b_{1.2})/c$	NUE (Ave + 10%)[5] $(a_{1.4}-b_{1.2})/c_{(ave+10\%)}$
	mt·ha⁻¹		kg·ha⁻¹	kg·kg⁻¹		%	
1982	10.7	7.9	106	100	59	51	30
1983	10.0	5.7	202	50	56	36	40
1984	12.2	6.8	202	61	68	44	50
1985	11.6	7.2	146	80	64	52	42
1986	13.5	7.5	157	86	75	63	55
Average	11.6	7.0	162	71	64	49	44

[1] Adapted from Bock and Hergert 1991.

[2] $N_{eff} \rightarrow$ kg grain produced per kg N applied at the optimum N rate.

[3] $N_{eff} \rightarrow$ kg grain produced per kg N applied at the average optimum N rate + 10% (180 kg N ha⁻¹).

[4] NUE $\rightarrow$ % fertilizer N recovered in the grain at the optimum N rate $\longrightarrow (a_{1.4}-b_{1.2})/c$ represents [(fertilized yield × 1.4%N − unfertilized yield × 1.2%N)/optimum N rate] × 100.

[5] NUE $\rightarrow$ % fertilizer N recovered in the grain at the average optimum N rate + 10% (180 kg N ha⁻¹) $\rightarrow (a_{1.4}-b_{1.2})/c_{(ave+10\%)}$ represents [(fertilized yield × 1.4% N − unfertilized yield × 1.2%N)/average optimum N rate + 10%] × 100.

Technologies for Predicting In-season N_{REQ}

Pre-Sidedress Soil Nitrate Test

If soil and environmental conditions result in a high probability of N_{inorg} being present at planting, then measuring N_{inorg} may enhance NUE. In coarse grains, assessing soil NO_3^- during early vegetative growth has been used to quantify in-season N rates (Magdoff et al. 1984). Presidedress soil nitrate test (PSNT) requires analysis of soil samples (0–30 cm) collected at V4 to V6 maize growth stage. Critical soil NO_3-N concentration below which N applications are recommended is (25 mg kg^{-1}, which varies between regions and crops. Semi-arid regions establish lower PSNT critical levels (13–15 mg kg^{-1}) because of greater N_{inorg}. Most studies show that PSNT explains 60 to 85 percent of the variation in crop response to in-season N (Sims et al. 1995).

With any N_{REQ} system based on soil and plant analysis, adoption is slow due to the labor and time required to collect, analyze, and interpret results, especially with only several weeks between soil sampling and in-season N application. Regions using PSNT have demonstrated 30 to 60 kg ha^{-1} reduced N rates and increased NUE.

Tissue Analysis

Petiole NO_3 is used to estimate in-season N applications and has resulted in increased yield and NUE in several crops. To assess N sufficiency in maize, NO_3 in stalk samples were collected after physiologic maturity (Fox et al. 2001). Similarly, grain protein can be used to indicate whether additional N is required for optimum yield. When winter and spring wheat grain protein is less than 11.5 and less than 13.2 percent, respectively, additional N would have increased yield (Goos et al. 1982).

Chlorophyll Meter

A chlorophyll meter measures the quantity of light (650 nm) transmitted through the leaf, where increasing chlorophyll content decreases light transmittance (Schepers et al. 1992). Leaf chlorophyll and percent N are highly correlated over the range of yield response to fertilizer N. Increasing N rate increases grain yield and leaf N, but chlorophyll readings do not increase with N applied above that required for optimum yield. In-season N is recommended when chlorophyll readings are less than 43, 44, and 37 for maize, wheat, and rice, respectively. Split N applications on rice after a chlorophyll assessment maintains yield potential with 12 to 25 percent less N and improves NUE 15 to 30 percent compared with pre-plant N (Singh et al. 2002).

A leaf color chart has been used instead of a chlorophyll meter with similar results (Balasubramanian et al., Chapter 2; Peoples et al., Chapter 4; Dobermann and Cassman, Chapter 19, this volume). In low-yield years, in-season chlorophyll monitoring

Table 12.2. Wheat grain yield response to nitrogen applied at uniform preplant and midseason (~ Feekes 6) rates compared with midseason nitrogen rates determined by remote sensing

N rate kg ha⁻¹	Method	Grain yield kg ha⁻¹	NUE %	Net revenue $ ha⁻¹
0		1182		118
45	Midseason	1562	25	131
90	Midseason	1810	17	132
90	½ Preplant + ½ Midseason	2105	22	161
90	Preplant	2063	22	157
43	Sensor NDVI / Midseason	1835	40	160
23	Sensor NDVI / ½ Midseason	1619	50	149

NUE, nitrogen use efficiency; NDVI, normalized difference vegetation index.

reduced N_{REQ} 34-101 kg ha⁻¹ with no yield loss compared with N_{REQ} based on pre-plant N_{inorg} (Chua et al. 2003). In high-yield years, in-season chlorophyll monitoring predicted higher N_{REQ}.

Remote Sensing

Remote sensing applications in agriculture have advanced rapidly, and many studies document the use of visible and near-infrared (NIR) spectral response from plant canopies to detect N stress (Ma et al. 1996). A strong correlation exists between crop N uptake and the normalized difference vegetation index (NDVI; Stone et al. 1996). NDVI is based on the principle that growing plants absorb visible light (photosynthetically active radiation, PAR), and reflect NIR radiation. NDVI is calculated as follows:

$$NDVI = \frac{NIR - PAR}{NIR + PAR}$$

Lukina et al. (2001) used late-tillering NDVI divided by growing degree days from planting to the time of measurement to predict in-season N_{REQ}. Raun et al. (2002) showed that prediction of wheat response to topdress N by remote sensing was positively correlated to measured N response and increased NUE and net return by 28 percent and 20 percent, respectively (Table 12.2).

Another method for estimating in-season N_{REQ} involves use of remote sensing to determine plant biomass (Weisz et al. 2001). Mid- and late-tillering are critical growth stages for N management to maximize wheat yield and NUE. If tiller density is less than 540 tillers m⁻², N application at GS25 improves grain yield. When density was greater than 540 tillers m^{-2}, N$_{REQ}$ is based on tissue N analysis. Tiller number is measured using aerial photography. Aerial color infrared photographs at GS25 also can be used to estimate optimum N rates applied at GS30.

Conclusions

With relatively fixed land resources to feed an increasing world population, significant advances in crop productivity per unit of land are essential to world food security. Over the last three decades, the dramatic increase in N use has translated into enhanced crop productivity but at significance risk to environmental quality. Minimizing the environmental impact of N use requires a substantial increase in NUE. Our current inability to estimate accurately the yield potential and the soil's capacity to supply N (N$_{OM}$) during a growing season represents the greatest challenge to increasing NUE.

Regardless of the methods used to estimate N$_{REQ}$, N applications synchronous with crop demand commonly increase NUE and reduce N$_{leach}$. Improving NUE requires technologies capable of quantifying plant N content early in the growing season and applying N at specific crop growth stages that ensures maximum crop recovery of applied N. Use of chlorophyll meters, leaf color charts, and aerial and ground-based remote sensing are valuable tools for measuring plant N, predicting crop yield potential, and estimating in-season N$_{REQ}$. Use of these technologies in conjunction with other geo-referenced field and soil information will enable producers to provide N in quantities predominately recovered by the crop. Further increases in NUE will come only from advances in technologies to quantify temporal and spatial distribution of N$_{OM}$ that will improve our ability to identify fertilizer N rates and application times synchronous with crop demand.

Literature Cited

Blevins, D. W., D. H. Wilkinson, B. P. Kelly, and S. R. Silva. 1996. Movement of nitrate fertilizer to glacial till and runoff from claypan soil. *Journal of Environmental Quality* 25:584–593.

Bock, B. R., and G. W. Hergert. 1991. Fertilizer nitrogen management. Pp.140–164 in *Managing nitrogen for groundwater quality and farm profitability,* edited by R. F. Follett et al. Madison, Wisconsin: Soil Science Society of America.

Chua, T. T., K. F. Bronson, J. D. Booker, J. W. Keeling, A. R. Mosier, J. P. Bordovsky, R. J. Lascano, C. J. Green, and E. Segarra. 2003. In-season nitrogen status sensing in irrigated cotton. I. Yields and nitrogen-15 recovery. *Soil Science Society of America Journal* 67:1428–1438.

Dabney, S. M., J. A. Delgado, and D. W. Reeves. 2001. Use of winter cover crops to improve soil and water quality. *Communications in Soil Science and Plant Analysis* 32:1221–1250.

FAO (Food and Agricultural Organization of the United Nations). 2001. *FAOSTAT Database collections.* http://www.apps.fao.org.

Firestone, M. K. 1982. Biological denitrification. Pp. 289–326 in *Nitrogen in agricultural soils,* edited by F. J. Stevenson. Madison, Wisconsin: ASA, CSSA, SSSA.

Fox, R. H., W. P. Piekielek, and K. E. Macneal. 2001. Comparison of late-seasons diagnostic tests for predicting nitrogen status of corn. *Agronomy Journal* 93:590–597.

Giller, K. E., G. Cadisch, C. Ehaliotis, E. Adams, W. Sakala, and P. Mafongoya. 1997. Building soil nitrogen capital in Africa. Pp.151–192 in *Replenishing soil fertility in Africa,* edited by R. J. Buresh et al. SSSA Spec. Publ. 51. Madison, Wisconsin: Soil Science Society of America and American Society of Agronomy.

Goos, R. J., D. G. Westfall, A. E. Ludwick, and J. E. Goris. 1982. Grain protein concentrations as an indicator of N sufficiency for winter wheat. *Agronomy Journal* 74:103–133.

Johnson, G. V., and W. R. Raun. 2003. Nitrogen response index as a guide to fertilizer management. *Journal of Plant Nutrition* 26:249–262.

Kanwar, R. S., T. S. Colvin, and D. L. Karlen. 1997. Ridge, moldboard, chisel, and no-till effects on tile water quality beneath two cropping systems. *Journal of Production Agriculture* 10:227–234.

Kurtz, L. T., L. V. Boone, T. R. Peck, and R. G. Hoeft. 1984. Crop rotations for efficient nitrogen use. Pp. 295–306 in *Nitrogen in crop production,* edited by R. D. Hauck. Madison, Wisconsin: American Society of Agronomy.

Lukina, E. V., K. W. Freeman, K. J. Wynn, W. E. Thomason, R. W. Mullen, G. V. Johnson, R. L. Elliott, M. L. Stone, J. B. Solie, and W. R. Raun. 2001. Nitrogen fertilization optimization algorithm based on in-season estimates of yield and plant nitrogen uptake. *Journal of Plant Nutrition* 24:885–898.

Ma, B. L., M. J. Morrison, and L. Dwyer. 1996. Canopy light reflectance and field greenness to assess nitrogen fertilization and yield of maize. *Agronomy Journal* 88:915–920.

Magdoff, F. R., D. Ross, and J. Amadon. 1984. A soil test for nitrogen availability to corn. *Soil Science Society of America Journal* 48:1301–1304.

Meisinger, J. J., and J. A. Delgado. 2002. Principles for managing nitrogen leaching. *Journal of Soil and Water Conservation* 57:485–498.

Meisinger, J. J., and G. W. Randall. 1991. Estimating nitrogen budgets for soil-crop systems. Pp. 85–124 in *Managing nitrogen for groundwater quality and farm profitability,* edited by R. F. Follett et al. Madison, Wisconsin: Soil Science Society of America.

Meisinger, J. J., F. R. Magdoff, and J. S. Schepers. 1992. Predicting N fertilizer needs for corn in humid regions: underlying principles. Pp. 7–27 in *Predicting N fertilizer needs for corn in humid regions,* edited by B. R. Bock and K. R. Kelling. Bull. Y-226. Muscle Shoals, Alabama: National Fertilizer Environmental Research Center.

Olson, R. A., W. R. Raun, Y. S. Chun, and J. Skopp. 1986. Nitrogen management and interseeding effects on irrigated corn and sorghum and on soil strength. *Agronomy Journal* 78:856–862.

Peoples, M. B., J. R. Freney, and A. R. Mosier. 1995. Minimizing gaseous losses of nitrogen. Pp.565–601 in *Nitrogen fertilization in the environment,* edited by P. E. Bacon. New York: Marcel Dekker.

Power, J. F., and F. E. Broadbent. 1989. Proper accounting for N in cropping systems. Pp.160–182 in *Managing nitrogen for groundwater quality and farm profitability,* edited by R. F. Follett, et al. Madison, Wisconsin: Soil Science Society of America.

Randall, G. W., and D. J. Mulla. 2002. Nitrate-N in surface waters as influenced by climate conditions and agricultural practices. *Journal of Environmental Quality* 30:337–344.

Raun, W. R., and G. V. Johnson. 1995. Soil-plant buffering of inorganic nitrogen in continuous winter wheat. *Agronomy Journal* 87:827–834.

Raun, W. R., J. B. Solie, G. V. Johnson, M. L. Stone, R. W. Mullen, K. W. Freeman, W. E. Thomason, and E. V. Lukina. 2002. Improving nitrogen use efficiency in cereal grain production with optical sensing and variable rate application. *Agronomy Journal* 94:815–820.

Rice, C. W., and J. L. Havlin. 1994. Integrating mineralizable nitrogen indices into fertilizer nitrogen recommendations. Pp.1–14 in *Soil testing: Prospects for improving nutrient recommendations,* edited by J. L. Havlin and J. S. Jacobsen. Special Publication 40. Madison, Wisconsin: Soil Science Society of America and American Society of Agronomy.

Rice, C. W., J. L. Havlin, and J. S. Schepers. 1995. Rational nitrogen fertilization in intensive cropping systems. *Fertilizer Research* 42:89–97.

Schepers, J. S., and R. H. Fox. 1989. Estimation of N budgets for crops. Pp. 221–246 in *Managing nitrogen for groundwater quality and farm profitability,* edited by R. F. Follett, et al. Madison, Wisconsin: Soil Science Society of America.

Schepers, J. S., and A. R. Mosier. 1991. Accounting for nitrogen in non-equilibrium soil-crop systems. Pp.125–138 in *Managing nitrogen for groundwater quality and farm profitability,* edited by R. F. Follett et al. Madison, Wisconsin: Soil Science Society of America.

Schepers, J. S., K. D. Frank, and C. Bourg. 1986. Effect of yield goal and residual soil nitrogen concentrations on N fertilizer recommendations for irrigated maize. *Journal of Fertilizer Issues* 3:133–139.

Schepers, J. S., T. M. Blackmer, and D. D. Francis. 1992. Predicting N fertilizer needs for corn in humid regions using chlorophyll meters. Pp. 103–114 in *Predicting N fertilizer needs for corn in humid regions,* edited by B .R. Bock and K. R. Kelling. Bull. Y-226. Muscle Shoals, Alabama: National Fertilizer Environmental Research Center.

Sims, J. T., B. L. Vasilas, K. L. Gartley. B. Milliken, and V. Green. 1995. Evaluation of soil and plant nitrogen tests for maize on manured soils of the Coastal Plain. *Agronomy Journal* 87:213–222.

Singh, B., Y. Singh, J. K. Ladha, K. F. Bronson, V. Balasubramanian, J. Singh, and C. S. Khind. 2002. Chlorophyll meter and leaf color chart based nitrogen management for rice and wheat in Northwestern India. *Agronomy Journal* 94:821–829.

Stone, M. L., J. B. Solie, W. R. Raun, R. W. Whitney, S. L. Taylor, and J .D. Ringer. 1996. Use of spectral radiance for correcting in-season fertilizer nitrogen deficiencies in winter wheat. *Transactions American Society of Agricultural Engineers* 39:1623–1631.

Tilman, D. 1999. Global environmental impacts of agricultural expansion: The need for sustainable and efficient practices. *Proceedings of the National Academy of Science* 96:5995–6000.

Vigil, M. F., B. Eghball, M. L. Cabrera, B. R. Jakubowski, and J. G. Davis. 2002. Accounting for seasonal nitrogen mineralization: A review. *Journal of Soil and Water Conservation* 57:464–469.

Weisz, R., C. R. Crozier, and R. W. Heiniger. 2001. Optimizing nitrogen application timing in no-till soft red winter wheat. *Agronomy Journal* 93:435–442.

PART V
Interactions and Scales

13

Fertilizer Nitrogen Use Efficiency as Influenced by Interactions with Other Nutrients

Milkha S. Aulakh and Sukhdev S. Malhi

Of the 16 essential plant nutrients, nitrogen (N) plays the most important role in augmenting agricultural production and affecting human and animal health. The amounts of different nutrients absorbed by a crop from soil may vary 10,000-fold, from 200 kg of N ha^{-1} to less than 20 g of Mo ha^{-1}, and yet rarely do these nutrients work in isolation. As agriculture becomes more intensive, the extent and severity of nutrient deficiencies and the practical significance of nutrient interactions increase. Interactions among nutrients occur when the supply of one nutrient affects the absorption, distribution, or function of another nutrient. In crop production, nutrient interactions assume added significance by affecting crop production and returns from investments made by farmers in fertilizers. Interaction between two or more nutrients can be positive *(synergistic)*, negative *(antagonistic)*, or even absent. When crop yield reaches an early plateau, this may be due to the limiting supply of another nutrient. When that nutrient is supplied, yield will continue to increase until another factor becomes limiting. When solar radiation, temperature, and soil water availability are non-limiting, plant nutrient requirements will be higher. When the need is fully satisfied for every factor involved in the process, the rate of the process can be at its maximum potential, which is greater than the sum of its parts because of sequentially additive interaction. Identification and exploitation of positive interactions hold the key to increasing returns in terms of yield, quality, and N use efficiency (NUE). Knowledge of the negative interactions is equally valuable because the test of precision crop nutrition lies in the ability to minimize the losses from antagonistic effects. In this chapter, we review and analyze the available information on the interaction of applied N with other nutrients on NUE.

N × P Interactions

It has frequently been shown that in a highly P-deficient soil, application of N alone has little impact on crop yield, but N + P application can dramatically increase the yield response to applied fertilizer (Table 13.1). The contribution of a synergistic interaction between N and P in cereals can be 13 to 89 percent of the yield response to N + P, depending on the yield potentials, the level of soil fertility, and nutrient application rates. If a soil is more deficient in P than N, then application of N alone could even cause a severe reduction in grain yield, as was observed by Sinha et al. (1973) in wheat (*Triticum aestivum* L.). Because application of P alone raised wheat grain yield by 682 kg ha^{-1}, the interaction impact was 79 percent on grain yield. In Vietnam, application of P reduced lodging and percentage of unfilled rice (*Oryza sativa* L.) grains resulting from the use of N alone and greatly improved yield response and NUE (Vo et al. 1995).

Several studies show that, in addition to enhanced crop yields, nutrient recoveries are higher in plots treated with N + P than with N or P alone. For example, grain response of sorghum (*Sorghum bicolor* L. Moench) per kg nutrient was higher by 11 percent when 120 kg nutrients ha^{-1} were distributed as 90 kg N + 30 kg P_2O_5 compared with only 120 kg N ha^{-1} (Sharma and Tandon 1992).

For non-irrigated crops, better root growth as a result of adequate P supply enables the plants to absorb water and nutrients from deeper layers during droughty spells, thereby increasing NUE. In early season maize (*Zea mays* L.) under dryland conditions of Bhagalpur, India, the N × P interaction was synergistic at all levels of N and P applied, but maximum interaction advantage was derived at 120 kg N + 60 kg P_2O_5 ha^{-1} (Singh 1991; Table 13.1). At this level, the interaction effect contributed 27 percent of the total yield response to N and P. Thus, the greater the investment in nutrients, the greater is the need for balanced nutrition.

Sunflower (*Helianthus annus* L.) yield is often increased by both N and P, but the interaction between the two nutrients may not be synergistic. Application of both N and P at lower rates, however, increased NUE by a factor of two (Aulakh and Pasricha 1996; Pasricha et al. 1987; Table 13.1).

The interaction of N × P in legumes, including grain legumes (e.g., pulses), oilseed legumes (e.g., peanut *Arachis hypogaea* L.), and soybean (*Glycine max* L. Merrill) is more complex because of biological N fixation (BNF). Under situations where the level of BNF is low, legumes may exhibit large responses to fertilizer N (Saimbhi and Grewal 1986). In their study, the yield of peas (*Pisum sativum* L.) increased by 2300 kg ha^{-1} or 70 percent with applied N, and the N × P interaction was synergistic, accounting for 14 percent of the N + P response. A positive N × P interaction may indicate poor BNF and greater dependence on fertilizer N. Application of P can create more favorable conditions for BNF. While application of N alone, particularly beyond 20 kg N ha^{-1}, reduced nitrogenase activity, balanced N and P application maintained nitrogenase activity at a high level in field pea (Pasricha et al. 1987).

Table 13.1. Influence of N × P interaction on nitrogen use efficiency (NUE) and apparent nitrogen recovery (ANR) in different field crops

Crop	Parameter	No fertilizer	N (kg N ha^{-1}) alone	P (kg P$_2$O$_5$ ha^{-1}) alone	N + P	Reference
Wheat	Grain yield (kg ha^{-1})	1750	4187 (120)	1947 (60)	5057	Dwivedi et al. 2003
	NUE (kg grain kg^{-1} N)		20.3		25.9	
	ANR (%)		45.5		55.3	
	Grain yield (kg ha^{-1})	1554	1270 (120)	2236 (90)	3473	Sinha et al. 1973
	NUE (kg grain kg^{-1} N)		-2.4		10.3	
Rice	Grain yield (kg ha^{-1})	2940	5530 (120)	3243 (60)	6190	Dwivedi et al. 2003
	NUE (kg grain kg^{-1} N)		21.6		24.6	
	ANR (%)		35.9		41.8	
Corn	Grain yield (kg ha^{-1})	1380	2440 (120)	1820 (60)	3450	Singh 1991
	NUE (kg grain kg^{-1} N)		8.8		13.6	
	Grain yield (kg ha^{-1})	1190	4750 (100)	2250 (60)	6750	Satyanarayana et al. 1978
	NUE (kg grain kg^{-1} N)		35.6		45.0	
Sorghum	Grain yield (kg ha^{-1})	2270	3670 (120)	3450 (60)	5500	Roy and Wright 1973
	NUE (kg grain kg^{-1} N)		11.7		17.1	
Sunflower	Seed yield (kg ha^{-1})	1470	1995 (60)	1672 (30)	2426	Aulakh and Pasricha 1996
	NUE (kg seed kg^{-1} N)		8.8		12.6	
Field pea	Grain yield (kg ha^{-1})	2180	2592 (40)	2422 (30)	3028	Pasricha et al. 1987
	NUE (kg grain kg^{-1} N)		10.3		15.2	

N × K Interactions

After N × P interaction, the N × K interaction is the second most important interaction in crop production. The significance of the N × K interaction and its optimum management are increasing as a result of increasing cropping intensity, higher crop yields, and greater depletion of soil K. Crops with a high requirement for K, such as maize and rice, often show strong N × K interaction.

Whereas the response of rice to P is more or less uniformly high at all levels of applied N, response to K increases with the amount of N + P applied (Umar et al. 1986). Increasing application rates from 40 kg N + 40 kg P_2O_5 ha^{-1} to 120 kg N + 140 kg P_2O_5 ha^{-1} increased rice yield by 300 and 960 kg ha^{-1} with 0 and 20 kg K_2O ha^{-1}, respectively. Application of NPK in the ratio of 120-40-0 and 40-40-20 produced similar rice yields, demonstrating higher nutrient use efficiency in the NPK treatment than with NP alone. Potassium increased rice yield by 250 kg ha^{-1} (7 percent) when N and P_2O_5 were applied at 40 kg ha^{-1} each but by 910 kg ha^{-1} (24 percent) at 120 kg N + 40 kg P_2O_5 ha^{-1}. Increasing N and P application rates without K application is often not a sound proposition and does not increase crop yield beyond a certain level. Higher levels of K are more effective at higher level of N and P. Other studies demonstrated that a weakly synergistic or additive N × P interaction could become highly synergistic when an adequate supply of K is ensured (Figure 13.1, a–c). In tropical soils, such as Ultisols and Oxisols, which are usually poor in available P and K, data from Brazil showed a positive N × K interaction in rice where a good response to K was obtained only when adequate N at 90 kg ha^{-1} was applied (PPI 1988, Figure 13.1d). Also, the response to N increased as the level of K was increased; the highest rice yield and NUE were obtained when both N and K were applied. Thus, it is clear that N × P × K interaction is helpful in increasing rice yields, provided N and P are applied in sufficient amounts.

N × S Interactions

The yield of wheat, grown in the coastal plain of Virginia, increased linearly with N + S application (Reneau et al. 1986). In four different field studies in India, application of S in addition to N and P produced additional yield of 700 to 1300 kg ha^{-1} for wheat and 400 kg ha^{-1} for corn (Aulakh and Chhibba 1992). In a field study, mixing urea with elemental S in a 4:1 ratio before its surface application onto a calcareous soil enhanced the NUE of pearl millets from 15 to 48 percent while reducing NH_3 volatilization by about 50 percent (Aggarwal et al. 1987).

Nitrogen and S are vital constituents of plant proteins and play a key role in oil production. When soils are deficient in available S, the yield of all crops is drastically reduced (Table 13.2). Oilseeds and legumes are more sensitive to S deficiency and more responsive to S fertilization than are cereals and grasses because of their higher

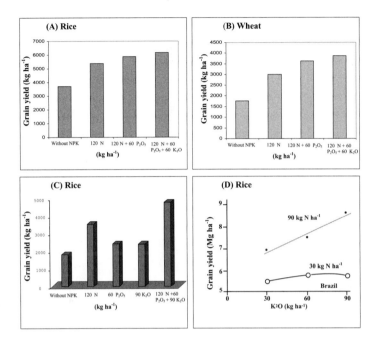

Figure 13.1. N × K or N × P × K interaction effects on (A) rice, (B) wheat (adapted from Singh and Bhandari 1995), (C) rice in India (prepared from Chandrakar et al. 1978), and (D) rice in Brazil (adapted from PPI 1988).

requirements for S. The quantity of S removed from soil for optimum crop yields is highest for oilseeds, followed by pulses and the lowest for cereals (Aulakh and Chhibba 1992). In a 3-year field study conducted on S-deficient Gray Luvisol soils in Saskatchewan, Canada, application of N fertilizer alone depressed yield and oil content of canola (*Brassica napus* L. or *Brassica rapa* L.), and NUE was –2.2 (Malhi and Gill 2002; Table 13.2). Compared with N alone, N + S fertilization increased yield from 140 kg seed ha^{-1} to 1228 kg seed ha^{-1}, and NUE from -2.2 kg seed kg N^{-1} to 3.7 kg seed kg N^{-1} (Table 13.2). McGrath and Zhao (1996) observed that without S application, the seed yield of *Brassica napus* declined drastically as a result of S deficiency when the N fertilization rate was increased from 180 to 230 kg ha^{-1}. Such severe negative impacts when N alone was applied to S-deficient soils on NUE, seed yield, oil content, and protein content in rapeseed and mustard crops were also observed in several other studies (Aulakh et al. 1980, 1995; Table 13.2). Apparent fertilizer N recovery in mustard seed increased from 25.1 to 39.6 percent and from about 65 to 80 percent in rapeseed (seed + straw) when N and S were applied together.

In forage crops, the higher yields generally obtained with N + S application suggest that the optimum ratios of N and S fertilizers must be determined for different soils and

Table 13.2. Influence of N × S interactions on nitrogen use efficiency (NUE) and apparent nitrogen recovery (ANR) in different field crops

Crop	Parameter	No fertilizer	N (kg N ha^{-1}) alone	S (kg S ha^{-1}) alone	N + S	Reference
Canola	Seed yield (kg ha^{-1})	406	140 (120)	779 (30)	1228	Malhi and Gill 2002
	NUE (kg seed kg^{-1} N)		−2.2		3.7	
Mustard	N uptake (kg N ha^{-1})	14.9	52.6 (150)	18.0 (60)	77.4	Aulakh et al. 1980
	ANR (%)		25.1		39.6	
Rapeseed	N uptake (kg N ha^{-1})	33.0	109.4 (150)	46.3 (20)	125.5	Aulakh et al. 1995
Grass	Dry matter yield (kg ha^{-1})	1410	940 (112)	1600 (11)	4640	Nyborg et al. 1999
	NUE (kg DM kg^{-1} N)		−4.2		27.1	
	N Uptake (kg N ha^{-1})		165 (200)		207	Brown et al. 2000
			278 (450)		332	

forages. In a 13-year field experiment on a Dark Gray Chernozem loam soil in Saskatchewan, the average NUE when N (112 kg ha^{-1}) was applied in isolation was –4.2 kg DM kg N^{-1} yr^{-1} (Nyborg et al. 1999; Table 13.2). The NUE increased to 27.1 kg DM kg N^{-1} yr^{-1} when S fertilizer was applied in combination with N.

Interactions of N with Ca, Mg, and Micronutrients

Although Ca requirements for plant growth and metabolism are low, it has great significance in balancing the levels of other nutrients, including N. In a highly acidic soil (pH 4.5), the substantially higher rice yields obtained with the combined application of lime and NPK than with lime or NPK alone indicated that soil acidity was the main constraint in the utilization of soil nutrients by the crop (Fageria and Baligar 2001). Once acidity was corrected, uptake of soil N increased many-fold. Other associated problems, such as high concentrations of Al and Mn and decreased root growth, may lead to a decline in NUE. In one of the experiments conducted by Malhi et al. (1995) in the Prairie Provinces of Canada, NUE of barley was increased by 20 and 12 kg grain kg N^{-1}, respectively, by the addition of lime, when N was applied at 50 and 100 kg N ha^{-1}. Grain yields were increased by 1 and 1.2 Mg ha^{-1}, respectively, by these treatments. These findings suggest that on acid soils, crops fertilized with N would show yield and NUE advantages from applications of lime.

Deficiencies of different micronutrients are not widespread, but whenever they occur, they can result in a serious reduction in grain yield and quality of crops and utilization efficiency of N. N and Zn show a synergistic effect, and best yields can be obtained with the optimum combination of both nutrients. In a field experiment on a sandy loam calcareous soil of Bihar, India, the addition of Zn with optimum N, P, and K increased NUE from 20.8 to 23.9 kg grain kg N^{-1} (Sakal et al. 1988).

Numerous reports suggest that N × Cu interaction can be synergistic (in soils with low Cu levels) or antagonistic (in soils when both nutrients are in excess supply). Cereals having protein-rich grains are more susceptible to Cu deficiency than are those poor in grain protein (Nambiar 1976).

Balanced Nutrition Globally

In addition to N losses (Goulding, Chapter 15, this volume), excessive N application can lead to a decline in crop production through deficiencies of macronutrients and micronutrients. The foregoing subsections have revealed that NUE could be improved with optimum and balanced use of different plant nutrients. Among these, most of the N, P, and K is supplied by synthetic fertilizers. In the period from 1960 through 1961 to and from 2001 through 2002, global N-fertilizer consumption increased from 10.8 Tg N yr^{-1} to 82.4 Tg N yr^{-1} (IFA 2003). The corresponding increase in the consumption of P and K fertilizers was from 4.7 Tg P yr^{-1} to 14.6 Tg P yr^{-1} and 1.1 Tg K yr^{-1}

to 19.1 Tg K yr^{-1}. There is no doubt that these fertilizers have contributed significantly to the continuing increase in grain production required to meet the demand of the increasing human and livestock population. The global distribution of fertilizers, however, has changed markedly in the past few decades. The use of N, P, and K fertilizers has declined in developed countries since 1985, but it has continued to increase linearly in the developing world over the past three decades. N is used in near-optimum amounts (or even in excessive amounts in some situations), whereas P and K are not always supplemented adequately (Aulakh and Bahl 2001; Mosier 2002).

Balanced and judicious use of fertilizers is the key to efficient nutrient use and for maintaining soil productivity. Balanced fertilization requires optimum input of N, P, and K in the ratio needed to maintain soil fertility and to optimize crop production. The main cereal crops, such as wheat, rice, and maize, typically have P:N ratios both in grain and straw in the narrow range of 0.15 to 0.24 (TFI 1982; Aulakh and Bahl 2001). Oilseeds such as sunflower, rapeseed, and linseed/flax (*Linum usitatisimum* L.) have similar P:N ratios in seed but much lower ratios in straw (0.07–0.10). The grains of legumes such as soybean, peanut, and mungbean have relatively low P:N ratios (0.05-0.12) because they accumulate high amounts of N through BNF. According to The Fertilizer Institute, if P and N fertilization is required, these should be applied in a P:N ratio of about 0.15 (TFI 1982).

The ratio of the global consumption of P:N in 1995 was 0.17 (0.39 P_2O_5/N), and K:N was 0.22 (0.26 K_2O/N), and these ratios have been predicted to remain relatively constant up to 2030 (Mosier 2002). A large disparity exists, however, in fertilizer consumption ratios within countries of a continent as well as among continents (FAO 2003). In North and Central America, the United States, and Canada, near-optimum amounts of N and P are being used, but K consumption is suboptimal in Canada. In South America, Brazil is using well above the optimal proportions of N, P, and K. In fact, the K:N ratio in Brazil is the highest in the world because fertilizer N is used in relatively small amounts for the predominantly grown soybean crop. On the other hand, all other countries in this continent may need to enhance the use of K fertilizers.

European countries show fewer variations in P:N and K:N ratios; they are quite close to desirable levels except in Germany and Russia, where the P:N and K:N ratios are very low, at 0.08 and 0.16, respectively. Within Asia, China and India are the highest consumers of fertilizers in the world, but they probably need to use substantially more K. Malaysia ranks second in K:N ratios (1.03), and Pakistan has the lowest K:N ratio (0.008) in the world. The P:N ratios in different regions/countries vary 10-fold (0.083–0.826), whereas the K:N ratios vary 150-fold (0.008–1.20).

Conclusions

The synergistic N × P interaction is responsible for a sizeable increase in yield gain, leading to considerable improvements both in N and P use efficiencies. The magnitude of this interaction is modified by soil type, level of available soil P, applied N and P rates, crop type, and climatic conditions. The overall trend of N × P interaction studies emphasizes the point that crop responses to N alone level off earlier, whereas those to N + P enable the crop to produce higher yields. Strongly positive and profitable interactions are possible in crops that have a high K requirement, and significant N × K interaction can be expected wherever higher doses of N are used to increase crop production.

Adequate N and S nutrition during plant growth is highly desirable for economic and stable production, and their application at optimum rates is required to improve the efficiency of each nutrient not only for crop yields but also for protein, oil production, and fatty acid quality. Correct diagnosis of nutrient deficiency is vital. If S deficiency is misdiagnosed as N deficiency and additional N is applied as a consequence, then crop growth would be adversely affected and a greater penalty would result in terms of crop yield and quality and NUE. Information on the N × Ca and N × Mg interactions is scanty and related mainly to the positive effects of lime in acidic soils and gypsum in sodic or solonetzic soils for correcting soil pH and improving plant growth.

Collectively, the benefits of improved NUE and other nutrients achieved by their balanced and optimum use include (1) a reduction in the amount of N used, resulting in lower costs to farmers; (2) realizing high-yield potentials as a result of synergistic nutrient interactions; (3) enabling the plant to resist damage from pests and diseases; (4) improving crop quality and biochemical constituents of the produce (e.g., protein, oil, fatty acids, nitrate); and (5) minimizing the amount of fertilizer nutrients left in the soil after harvest, thus reducing the potential for negative environmental impacts.

Literature Cited

Aggarwal, R. K., P. Raina, and P. Kumar. 1987. Ammonia volatilization losses from urea and their possible management for increasing nitrogen use efficiency in an arid region. *Journal of Arid Environments* 13:163–168.

Aulakh, M. S., and G. S. Bahl. 2001. Nutrient mining in agro-climatic zones of Punjab. *Fertiliser News* 46:47–61.

Aulakh, M. S., and I. M. Chhibba. 1992. Sulphur in soils and responses of crops to its application in Punjab. *Fertiliser News* 37:33–45.

Aulakh, M. S., and N. S. Pasricha. 1996. Nitrogen and phosphorus requirement and ability to scavenge soil N by hybrid sunflower. *Crop Improvement* 23:247–252.

Aulakh, M. S., N. S. Pasricha, and K. L. Ahuja. 1995. Effect of nitrogen and sulphur application on grain and oil yield, nutrient uptake and protein content in transplanted *gobhi sarson* (*Brassica napus* L. subsp *oleifera* var *annua*). *Indian Journal of Agricultural Science* 65:478–482.

Aulakh, M. S., N. S. Pasricha, and N. S. Sahota. 1980. Yield, nutrient concentration and quality of mustard crops as influenced by nitrogen and sulphur fertilizers. *Journal of Agricultural Science, Cambridge* 94:545–549.

Brown, L., D. Scholefield, E. C. Jewkes, N. Preedy, K. Wadge, and M. Butler. 2000. The effect of sulphur application on the efficiency of nitrogen use in two contrasting grass-land soils. *Journal of Agricultural Science, Cambridge* 135:131–138.

Chandrakar, B. L., R. A. Khan, D. V. S. Chavhan, and B. P. Dubey. 1978. Response of dwarf rice to potash application under different nitrogen and phosphorus combinations. *Indian Potash Journal* 3:21–23.

Dwivedi, B. S., A. K. Shukla, V. K. Singh, and R. L. Yadav. 2003. Improving nitrogen and phosphorus use efficiencies through inclusion of forage cowpea in the rice-wheat systems in the Indo-Gangetic plains of India. *Field Crops Research* 80:167–193.

Fageria, N. K., and V. C. Baligar. 2001. Improving nutrient use efficiency of annual crops in Brazilian acid soils for sustainable crop production. *Communications in Soil Science and Plant Analyses* 32:1303–1319.

FAO (Food and Agricultural Organization). 2003. *FAOSTAT: Agricultural Data* (http://fao.org). Rome, Italy: Food and Agricultural Organization.

IFA (International Fertilizer Industry Association). 2003. *Data statistics.* (http://www.fertilizer.org). Paris, France: International Fertilizer Industry Association.

Malhi, S. S., and K. S. Gill. 2002. Effectiveness of sulphate-S fertilization at different growth stages for yield, seed quality and S uptake of canola. *Canadian Journal of Plant Science* 82:665–674.

Malhi, S. S., G. Mumey, M. Nyborg, H. Ukrainetz, and D. C. Penney. 1995. Longevity of liming in western Canada: Soil pH, crop yield and economics. Pp. 703–710 in *Plant–soil interactions at low pH*, edited by R. A. Date, N. J. Grundon, G. E. Rayment, and M. E. Probert. Dordrecht, The Netherlands: Kluwer Academic Publishers.

McGrath, S. P., and F. J. Zhao. 1996. Sulphur uptake, yield response and the interactions between N and S in winter oilseed rape (*Brassica napus*). *Journal of Agricultural Science, Cambridge* 126:53–62.

Mosier, A. R. 2002. Environmental challenges associated with needed increases in global nitrogen fixation. *Nutrient Cycling in Agroecosystems* 63:101–116.

Nambiar, E. K. S. 1976. Genetic differences in the copper nutrition of cereals. 2. Genotypic differences in response to copper in relation to copper, nitrogen and mineral contents of plants. *Australian Journal of Agricultural Research* 27:453–463.

Nyborg, M., S. S. Malhi, E. D. Solberg, and R. C. Izaurralde. 1999. Carbon storage and light fraction C in grassland Dark Gray Chernozem soil as influenced by N and S fertilization. *Canadian Journal of Soil Science* 79:317–320.

Pasricha, N. S., M. S. Aulakh, G. S. Bahl, and H. S. Baddesha. 1987. *Nutritional requirements of oilseed and pulse crops in Punjab (1976–1986).* Research Bulletin 15. Ludhiana, India: Department of Soils, Punjab Agricultural University.

PPI (Phosphate and Potash Institute). 1988. Effects of N and K fertilization in rice crop. *Better Crops International* December 1988, 9.

Reneau, R. B. Jr., D. E. Bran, and S. J. Donohue. 1986. Effect of sulphur on winter wheat grown in the coastal plain of Virginia. *Communications in Soil Science and Plant Analysis* 17:149–158.

Roy, R. N., and B. C. Wright. 1973. Sorghum growth and nutrient uptake in relation to soil fertility. I. Drymatter accumulation pattern, yield and N content of grain. *Agronomy Journal* 65:709–711.

Saimbhi, M. S., and A. S. Grewal. 1986. Effect of sources of N and levels of N and P on the growth, nutrient uptake and yield of pea. *Punjab Vegetable Grower* 21:10–15.

Sakal, R., A. P. Singh, and R. B. Sinha. 1988. Effect of different soil fertility levels on response of wheat to zinc application on calciorthent. *Journal of Indian Society of Soil Science* 36:125–127.

Satyanarayana, T., V. P. Badanur, and G. V. Havanagi. 1978. Response of maize to nitrogen, phosphorus and potassium on acid sandy loam soils of Bangalore. *Indian Journal of Agronomy* 23:49–51.

Sharma P. K., and H. L. S. Tandon. 1992. The interaction between nitrogen and phosphorus in crop production. Pp. 1–20 in *Management of nutrient interactions in agriculture*, edited by H. L. S. Tandon. New Delhi, India: Fertiliser Development and Consultation Organisation.

Singh, A. K. 1991. Response of pre-flood, early rainy-season maize (*Zea mays*) to graded levels of nitrogen and phosphorus in Ganga *diara* tract of Bihar. *Indian Journal of Agronomy* 36:508–510.

Singh, B., and A. L. Bhandari. 1995. Response of cereals to applied potassium in Punjab. Pp. 58–68 in *Use of potassium in Punjab agriculture*, edited by G. Dev and P. S. Sidhu. Gurgaon, Haryana, India: Potash and Phosphate Institute of Canada— India Programme.

Sinha, M. N., A. G. Kavitkar, and M. Parshad. 1973. Optimum nitrogen and phosphorus requirements of late-sown wheat (*Triticum aestivum* L.). *Indian Journal of Agricultural Science* 43:1002–1005.

TFI (The Fertilizer Institute). 1982. *The fertilizer handbook.* Washington, DC: TFI.

Umar, S. M., B. Prasad, and B. Prasad. 1986. Response of rice to N, P and K in relation to soil fertility. *Journal of Indian Society of Soil Science* 34:622–624.

Vo, T. G., T. L. Tran, M. H. Nguyen, E. G. Castillo, J. L. Padilla, and U. Singh. 1995. Nitrogen use efficiency in direct-seeded rice in the Mekong River Delta, Vietnam: Varietal and phosphorus response. Pp. 151–159 in *Proceedings of a conference on Vietnam and IRRI partnership in rice research, Hanoi (Vietnam)*, edited by G. L. Denning and T. X. Vo. Los Banos, Laguna, Philippines: International Rice Research Institute, and Hanoi, Vietnam: Ministry of Agriculture and Food Industry.

14

An Assessment of Fertilizer Nitrogen Recovery Efficiency by Grain Crops

T. J. Krupnik, J. Six, J. K. Ladha, M. J. Paine, and C. van Kessel

The increased N pollution of waters and the atmosphere and the predicted further increase of the global population underscore the pressing need for improving N use efficiency (NUE) in crop production (Janzen et al. 2003). This can be accomplished only when this effort is based on a thorough quantification and understanding of the factors determining fertilizer N recovery efficiency. We conducted a literature review to estimate fertilizer N recovery efficiency at a variety of scales, ranging from the research plot/farmer's field, to the farm, to the region (at the farm and regional scale, total N input rather than fertilizer input was considered). Studies applying the N balance or the ^{15}N isotope dilution method were considered to assess fertilizer N recovery efficiency. Recoveries were based on grain N and grain plus straw N. Only wheat, maize, and rice were included in our analysis because we did not find sufficient data for other crops across all the regions of the world.

Data Collection

We selected data points on fertilizer N recovery efficiency from 175 field studies conducted across all regions of the world. Efforts were made to include field studies that reported both the N difference and the ^{15}N method for calculating fertilizer N recovery efficiency. Only data published in peer-reviewed journals were considered. The complete list of included publications can be found at http://agronomy.ucdavis /vankessel/NE. If one particular field study reported data for multiple seasons or when similar fertilizer N uptake trials were conducted at the same site using different varieties or sources of fertilizer N, individual entries for fertilizer N recovery were made for the different seasons, varieties, or sources of N fertilizer. At the farm and regional scale, only data from crop production systems without livestock components were used.

Defining Fertilizer Nitrogen Recovery Efficiency

The fertilizer N recovery efficiency is commonly calculated by the N difference method, also referred to as the *N balance* or the *apparent recovery efficiency* of applied fertilizer N method (RE_N). This method requires a zero-N fertilizer control plot. Fertilizer N recovery can also be determined by the ^{15}N isotope dilution (RE_{15N}) method. Labeled ^{15}N fertilizer is applied and its recovery in the crop determined. For the ^{15}N method, a zero-N fertilized control plot is not required. For both methods, the recovery measurements can be based on total N in the grain or on total N in grain plus straw.

At the farm or regional level, it becomes increasingly difficult or near impossible to install a sufficiently large number of zero-N fertilized plots or to use ^{15}N-labeled fertilizers. At this scale, overall RE rather than RE_N is reported. In addition to fertilizer N, input of N includes biological fixed N_2, atmospheric deposition, N in irrigation water, animal manures, and residue N (Janzen et al. 2003). Losses of N include leaching, gaseous emissions from volatilization and denitrification, and losses via erosion when soil particles are exported beyond the study area. The RE is based on a total N budget of N input/N losses compared with the amount of N accumulated in the crop (Frissel 1978); however, RE of all the different sources of N is not equal, and therefore no accurate estimate of the RE_N can be made (Cassman et al. 2002). The RE_N of a system is assumed to become equal to RE once the system is near steady-state (Cassman et al. 2002).

Fertilizer Nitrogen Recovery at the Research Plot Level

Across all regions, crops, and methods, a wide (5–96 percent) range of estimates for RE_N and RE_{15N} in the grain was observed (Table 14.1).

Across all regions, the average recovery of N fertilizer for the three crops was the lowest for Africa (i.e., 26 percent for the RE_N method and 22 percent by the RE_{15N} method). These lower values may not be surprising because of other prevalent growth-limiting factors, such as a water or P deficiency found in much of Africa. The highest regional recovery was observed for South America with the RE_N method: 52 percent. A much lower average recovery was found, however, when the RE_{15N} method was used: 33 percent (Table 14.1). Some caution in the interpretation of the higher value for RE_N for South America is warranted here because the number of observations was low.

The data set represents an approach to calculating RE_N at large scales. When examined globally by crop type, grain RE_N was highest for maize (39 percent), followed by wheat (38 percent) and lowest for rice (36 percent). When based on the uptake of ^{15}N fertilizers, similar values were found for maize and wheat (37 percent), with rice showing a recovery of 32 percent. The lower recoveries of N fertilizers by rice may be caused by the anaerobic soil conditions of paddy rice and the reduced form of the N fertilizer applied, which can lead to higher N losses via NH_3 volatilization and denitrification following the conversion of NH_4^+ to NO_3^-.

Table 14.1. Recovery of nitrogen fertilizer in grain by maize, rice, and wheat across regions of the world determined by RE_N and RE_{15N} methods

	Fertilizer N recovery (%) RE_N method					Fertilizer N recovery (%) RE_{15N} method				
	N fertilizer rate (kg ha^{-1})	Mean	Maximum	Minimum	∂n	N fertilizer rate (kg ha^{-1})	Mean	Maximum	Minimum	∂n
Africa										
Maize	—	—	—	—	—	68	23	41	5	25
Rice	124	24	41	10	47	—	—	—	6	3
Wheat	138	49	59	39	4	90	13	18	5	3
Averages/totals	121	26	59	10	51	79	22	41	5	28
Australia										
Maize	—	—	—	—	—	—	—	—	—	5
Rice	175	32	36	29	6	120	25	38	15	5
Wheat	89	38	77	7	42	79	28	45	6	43
Averages/totals	132	37	77	7	48	99	28	45	6	48
Eurasia										
Maize	—	—	—	—	—	—	—	—	—	32
Rice	115	41	54	32	3	68	32	73	7	32
Wheat	119	27	53	7	7	168	47	89	38	40
Averages/totals	117	31	54	7	10	118	40	89	7	72
Europe										
Maize	—	—	—	—	—	—	—	—	—	—
Rice	—	—	—	—	—	—	—	—	—	—
Wheat	156	43	87	6	78	135	41	65	16	106
Averages/totals	156	43	87	6	78	135	41	65	16	106

(continued on page 196)

Table 14.1 *(Continued)*. Recovery of nitrogen fertilizer in grain by maize, rice, and wheat across regions of the world determined by RE_N and RE_{15N} methods

	Fertilizer N recovery (%) RE_N method					Fertilizer N recovery (%) RE_{15N} method				
	N fertilizer rate (kg ha⁻¹)	Mean	Maximum	Minimum	∂n	N fertilizer rate (kg ha⁻¹)	Mean	Maximum	Minimum	∂n
North America										
Maize	139	39	71	8	46	160	39	87	7	128
Rice	—	—	—	—	—	39	28	52	14	12
Wheat	91	35	87	6	222	89	35	94	5	152
Averages/totals	115	36	87	6	268	96	34	94	7	292
South America										
Maize	240	31	40	27	3	240	48	65	45	6
Rice	120	39	50	32	9	—	—	—	—	—
Wheat	126	69	86	24	10	131	17	46	11	6
Averages/totals	162	52	86	24	22	186	33	65	11	12
South Asia										
Maize	80	30	41	23	3	—	—	—	—	—
Rice	213	39	93	7	213	121	32	96	5	196
Wheat	55	49	67	24	55	131	39	83	24	50
Averages/totals	116	41	93	7	271	126	33	96	5	246
Totals/averages by crop										
Maize	153	38	71	8	52	156	37	87	5	159
Rice	149	36	93	7	278	87	32	96	5	245
Wheat	111	39	87	6	418	118	37	94	6	400
Averages/totals for all regions		38			748		35			804

∂n, number of observations; RE, recovery efficiency.

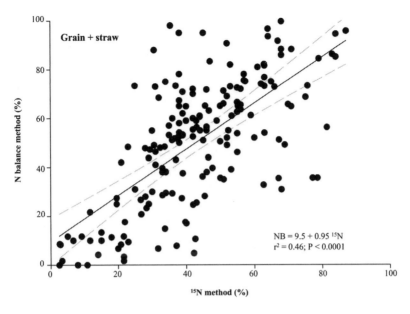

Figure 14.1 Relationships between recovery efficiency of nitrogen (RE_N) and RE_{15N} for measuring the efficiency of fertilizer N recovery in grain and grain plus straw of major cereal crops.

When averaged for all crops and regions, the overall recovery of fertilizer N based on the RE_N was 38 percent and reduced slightly to 35 percent when based on the NE_{15N} method (Table 14.1). Lower recoveries of fertilizer N by crops based on ^{15}N-labeled fertilizer are often observed (see below).

The wide but consistent range in estimates indicates that we have appropriate methodologies to assess RE_N but that RE_N is affected by a multitude of factors of which we currently have limited understanding. Nevertheless, it is generally accepted that RE_N will decrease with increasing fertilizer N rates because of increased chances for N losses through runoff, leaching, and gaseous emissions (Baligar et al. 2001). In addition, Pilbeam (1996) found that the percentage of RE_{15N} across locations is strongly related to the precipitation–evaporation quotient. Interestingly, our average RE_N values were generally higher than those calculated by Raun and Johnson (1999), which were calculated on the basis of total N applied and harvested on a global basis. Additionally, they assumed grain N concentrations to derive crop N uptake. This resulted in a rather crude estimate of global grain RE_N of 33 percent. Fertilizer N recovery increased further by 9 percent for both methods (RE_N method: from 38 to 47 percent; RE_{15N} method: from 35 to 44 percent) when straw was included (data not shown). Straw remains an important sink for fertilizer N and following incorporation, a source of fertilizer N for the subsequent crops.

Roberts and Janzen (1990) showed that estimates of RE_{15N} and RE_N are statistically related, although not as closely as might be expected (r^2 = 46 percent; Figure 14.1); however, the slopes and intercepts of the regression differ from unity and zero, respectively. Consequently, the two methods overestimate or underestimate fertilizer N recovery compared with each other. Our data set suggests that the ^{15}N dilution method provides higher recoveries of fertilizer N for grain at the high range of RE_N and underestimates fertilizer N recovery at its lower range compared with the N balance method (Figure 14.1). The opposite was observed, however, when the recovery in grain plus straw was included. Nevertheless, the average RE_N across all regions (Table 14.1) for grain was 3 percent lower (35 versus 38 percent) when estimated by the ^{15}N dilution method than by the N balance method. Others have also reported generally lower estimates for the ^{15}N dilution method (Roberts and Janzen 1990). The initial immobilization of ^{15}N fertilizer in the microbial biomass and the early release of ^{14}N from the microbial biomass are likely the causes of a lower ^{15}N-fertilizer recovery in the crop.

Smil (1999, 2002) argued that the ^{15}N dilution method provides a more accurate estimate of fertilizer NUE than the N balance method. In contrast, Cassman et al. (2002) consider the N balance method more reliable as the N balance method is influenced by fewer confounding factors. As pointed out by Cassman et al. (2002), both methods can lead to erroneous estimates. For the N balance method, estimates can be confounded by so-called added N interactions. If the addition of N fertilizer leads to an increase in crop available N, an overestimation of fertilizer N use efficiency will occur. This increase in crop available N can occur if root development of N-fertilized crop increases (accumulating N from deeper depth compared with unfertilized crops) or if the rate of net N mineralization of soil organic matter or residues increases with the addition of N fertilizers. With respect to the ^{15}N-isotope method, added N effects can lead to an underestimation of the amount of ^{15}N fertilizer accumulated by the crop. The main cause for the underestimation is pool N substitution, which causes immobilization of ^{15}N fertilizer in the microbial biomass and the initial release of microbial-derived ^{14}N. Although the difference in the REN estimates might appear small, the implications of these small differences cannot be ignored. For example, a 1 percent increase in REN was calculated to save approximately $234 million (U.S. dollars) worldwide in fertilizer costs.

Fertilizer Nitrogen Recoveries in Subsequent Crops

Whereas the vast majority of published data on fertilizer N recoveries are based on the first growing season (Table 14.1), fewer studies have investigated the residual effect of fertilizer N uptake by subsequent crops. Quantifying the amount of fertilizer N that remains available for the subsequent crops can be assessed only with the use of labeled ^{15}N fertilizer.

The ^{15}N fertilizer accumulated by the crops in subsequent years is likely the result of net mineralization of crop residues or microbial biomass rather than unused fertilizer N that remained as inorganic soil since the fertilizer was applied (Hart et al. 1993). Plant-available ^{15}N fertilizer can include N from above as well as belowground plant residues.

The International Atomic Energy Agency (IAEA) initiated a comprehensive and long-term collaborative research program with the main objective being determination of residual fertilizer ^{15}N uptake by a variety of crops during subsequent growing seasons as influenced by residue management (IAEA 2003). Experiments were conducted at 13 locations across Africa, Asia, and South America. The recovery in the crop and soil of a single application of ^{15}N fertilizer was determined for up to six growing seasons. Crop residues were removed or incorporated. When residues were incorporated or removed, the average percentage of the single application of ^{15}N fertilizer recovered in the above-ground residue and grain in the subsequent five growing seasons, that is, excluding the first growing season, across all locations was 7.1 percent and 5.7 percent, respectively. Total recovery of ^{15}N fertilizer in the crop plus soil averaged 65 percent in the first growing season when the residues were not removed and 66 percent when they were removed. At the end of the sixth growing season, combined total ^{15}N fertilizer recovery in the crops and soil had decreased to 58 percent when residues were incorporated and to 57 percent when residues were removed. The average amount of ^{15}N fertilizer recovered in soil after five growing seasons across all sites was 15 percent. Whether crop residues were removed or incorporated, no differences in total ^{15}N fertilizer recoveries by the subsequent crops were observed.

The residual ^{15}N fertilizer recoveries by the subsequent crop across a diverse range of cropping systems have been reported (Table 14.2). In a limited number of studies, the uptake of residual fertilizer was followed for several growing seasons (IAEA 2003). Cropping systems included flooded (rice) and dryland systems, and the amount of ^{15}N-fertilizer applied in the first year ranged between 30 and 196 kg N ha^{-1}. The forms of N applied included $(NH_4)_2SO_4$, urea, or NH_4NO_3.

The recovery of ^{15}N fertilizer in the first subsequent crop ranged from a minimum of 1.9 percent for a wheat–wheat system to a maximum of 5 percent for a rice–rice system (Table 14.2). The average recovery of ^{15}N fertilizer in the first subsequent crop across all systems (a total of 72 independent measurements) was 3.3 percent of the applied N (Table 14.2). The average ^{15}N fertilizer recovery was 1.3 percent for the second subsequent crop, 1.0 percent for the third subsequent crop, 0.4 percent for the fourth subsequent crop, and 0.5 percent for the fifth subsequent crop. Neither the form nor the amount of ^{15}N fertilizer applied nor the crop tested had a significant effect on the recovery of fertilizer N by the subsequent crops.

The average accumulated recovery of ^{15}N fertilizer by the subsequent crops during five growing seasons amounted to 6.5 percent, which is equal to 16 percent of the total fertilizer N recovered during the first growing season. With our calculated average, fertilizer ^{15}N recovery of 44 percent in grain and straw in the first growing season, the additional uptake by the five subsequent crops brings the total recovery to about 50 percent. Assuming that the amount of ^{15}N in the roots becomes negligible in the sixth growing season, the remaining 50 percent of the ^{15}N fertilizer would have become part of (1) the soil organic matter pool (with potential for later crop uptake) or (2) lost from the cropping system entirely (Jansson and Persson 1982).

In general, most ^{15}N fertilizer is lost during the year of application (IAEA 2003).

Table 14.2. Residual ^{15}N fertilizer recovery by subsequent crops at different rates of applied nitrogen

First crop	Subsequent crop	Source (kg ^{15}N)	^{15}N Fertilizer recovered (%)	Comments
Rice (wet season)	Rice	Urea (87 kg ha^{-1})	2.4	Average of 5 application methods
Rice (dry season)	Rice	Urea (58 kg ha^{-1})	3.4	Average of 5 application methods
Wheat	Sunflower	AS (100 kg ha^{-1})	3.6	Split applied
Rye	Sugarbeet	AS (100 kg ha^{-1})	2.0	Average of 2 times of application
Rice	Rice	Urea (54 kg ha^{-1})	5.0	Average of 4 management practices
Rice	Rice	Urea (70 kg ha^{-1})	4.8	Average of 3 management practices
Rice	Rice	Urea (60–120 kg ha^{-1})	1.5	Average of 4 management practices
Wheat	Wheat	AN (47–196 kg ha^{-1})	2.0	Average of 11 sites
			1.0[1]	Average of 7 sites
			0.7[2]	Average of 6 sites
			0.7[3]	Average of 2 sites
Rice	Rice	urea (20 kg ha^{-1})	3.0	Average of 4 management practices
Wheat	Wheat	AS (120 kg ha^{-1})	3.2	Average of 4 residue practices
			2.1[1]	
			1.2[1]	
Ryegrass	Wheat	AS (120 kg ha^{-1})	4.2	Average of 4 residue practices
			1.9[1]	
			1.4[2]	

Oats	Wheat	AN (100 kg ha^{-1})	4.7	Average of 10 straw-tillage management practices
Wheat	Wheat	urea/AS/KNO$_3$ (25–75 kg ha^{-1})	3.1	Average of 3 sources of N at 2 rates
Wheat	Wheat	urea (134 kg ha^{-1})	1.9	Average of 3 times of N applications
Various crops	Various crops	AS (30–60 kg ha^{-1})	3.9	Average of 13 sites
			1.0[1]	Average of 13 sites
			0.7[2]	Average of 7 sites
			0.3[3]	Average of 6 sites
			0.5[4]	Average of 4 sites
All crops and sites (weighed average)			3.3	1st subsequent crop
			1.3	2nd subsequent crop
			1.0	3rd subsequent crop
			0.4	4th subsequent crop
			0.5	5th subsequent crop

[1] Second subsequent crop.
[2] Third subsequent crop.
[3] Fourth subsequent crop.
[4] Fifth subsequent crop.
AS, Ammonium sulfate; AN, Ammonium nitrate.

Although residual N fertilizer will serve as only a minor source of N to meet the crop's demand for N, its cumulative effect during the subsequent growing seasons should not be ignored when management decisions on long-term strategies to increase fertilizer N recovery efficiencies are developed.

Management decisions to increase fertilizer N use by crops can be focused on two strategic approaches: (1) to increase fertilizer N use during the first growing season when the fertilizer is applied or (2) to decrease N fertilizer loss thereby increasing the potential recovery of residual N fertilizer by the subsequent crops. Removing plant growth–limiting factors would increase the demand for N by the crop leading to a higher use of available N and, consequently, higher NUE (Balasubramanian et al., Chapter 2, this volume). Fully synchronizing N fertilizer application with crop N demand will lead to higher N fertilizer use efficiency. In other words, "get the right nutrients in the right amount at the right time at the right place" (Oenema and Pietrzak 2002). On the other hand, management practices focused on reducing N fertilizer losses from a cropping system may not always lead to higher fertilizer NUE during the first growing season but can lead to an increase in the recovery of fertilizer N by subsequent crops. Management practices, which are focused on increasing fertilizer NUE as measured over a number of growing seasons instead of the first growing season when the fertilizer is applied, have received limited attention and remain largely untested.

Fertilizer Nitrogen Recovery Under Farm Conditions

In contrast to the preceding data, which come exclusively from manipulative experiments conducted at field research stations, we also considered studies of RE_N determined only under on-farm conditions (Dobermann et al. 2002, 2004; Haefele et al. 2003). As expected, the average RE_N estimates from on-farm assessments are lower than the average reported RE_N values determined at research stations, especially for maize and wheat. This discrepancy in estimates is due to the different scale of farming practices (Cassman et al. 2002). Experimental farms and research stations are more intensively managed than farmers' fields. The larger scale of on-farm experiments leads to a higher spatial variability of factors controlling RE_N, less stringent and suboptimal management, and decreased ability to exercise precise and detailed observations. The smaller difference between on-farm and research station estimates of RE_N for rice compared with RE_N for wheat and maize might be a result of the smaller difference in scale between research plots and farmers fields for rice than for wheat and maize. In addition, rice is generally more intensively managed than wheat or maize.

Nitrogen Recovery at the Farm and Regional Level

Using the RE_N and calculating net N inputs and outputs, N recovery by crops and total N losses of all combined N inputs can be calculated (Table 14.3). To estimate

Table 14.3. Input and uptake of nitrogen by crops and nitrogen recovery efficiency at the farm and regional scale

	Input				Recovery		
	Fertilizer	N_2 fixation	NO_x deposition	Other[1]	Total	Crop uptake	%	Reference
Arable farms [2] (kg N ha^{-1})	219	5.0	9.0	58.0	285.0	179[3]	73	Frissel 1978
Country/region (Tg N)								
United States	11	5.9	1.4	—	18.5	10.5[4]	56	Howarth et al. 2002
Canada	2	0.4	0.3	—	2.4	1.23[5]	52	Janzen et al. 2003
World	78	33.0	20.0	38.0	169.0	85.0	50	Smil 1999
World	78	7.7	21.6	68.9	176.4	101.2	57	Sheldrick et al. 2002

[1] Includes seeds, irrigation water, crop residues and animal manures.
[2] Average of 7 arable farms.
[3] Does not include residue and root-N.
[4] Above and belowground biomass.
[5] Includes 0.2 Tg in animal products.

N recovery by the crops or to determine N losses, the internal cycling of N from soil organic matter is not included (Janzen et al. 2003). Total crop N accumulation is instead based on N accumulation in grain and straw (Smil 1999) or can also include root N (Janzen et al. 2003). As all forms of N input and output are included, the recovery of N by the crop is not a reflection anymore of fertilizer-N recovery but includes the recovery of all forms of N input such as wet and dry deposition and biological N_2 fixation.

Frissel (1978) reported a major work on nutrient cycling in agricultural ecosystems, which included seven intensive arable cropping systems in the European Union (EU), the United States, Israel, and South America. Changes in the amounts of N in plant, animal, and soil components were measured, and N input via fertilizers, manure/waste, irrigation water, and wet and dry deposition determined. None of the systems included here in this current review had an animal component and systems with a high N input via biological N_2 fixation were excluded. Average total annual N input was 285 kg N ha^{-1} of which the majority, 219 kg Nha^{-1}, was from fertilizer N (77 percent). Other inputs included N from crop residues, biological N_2 fixation, and N deposition. In these studies, total crop N uptake included above- and below-ground plant components.

More recently, total crop N recoveries across countries or the world have been reported (Table 14.3). Smil (1999) estimated that on the world scale 50 percent of all input N was recovered by the harvested crop and their residues. Sheldrick et al. (2002) calculated a slightly higher value of 57 percent. Similar values were reported for Canada (52 percent) and the United States (56 percent; Howarth et al. 2002). These values are also close to the values found when the recovery of ^{15}N fertilizer is followed in the crops for six growing seasons (Table 14.2).

Crop N recovery values for the United States, however, did not include N in residues and roots (Howarth et al. 2002). In contrast, in the Canadian study, N accumulated in residue, roots, and animal products were included. Janzen et al. (2003) calculated wheat dry matter allocation (grain:aboveground residue:root) to be 0.34:0.51:0.15. Assuming an N concentration of 3 percent in the grain, 0.7 percent in the residues, and 0.5 percent in the roots, 30 percent of the crop N will be in the roots and residue combined. Assuming that crops in the United States have, on average, a similar grain:residue:root ratio and N content in residue and root, total crop recovery of N in the United States would increase to 79 percent, a value closer to that observed for the arable cropping systems reported by Frissel (1978).

Quantifying losses of N from agricultural fields remain prone to large uncertainties, in particular losses via denitrification and volatilization. Fortunately, once a cropping system is in near steady state with respect to its N content, and inputs via wet and dry deposition and biological N_2 fixation are considered constant, it can be argued with sufficient confidence that a total N budget should provide a good indicator of N fertilizer recovery efficiency (Cassman et al. 2002).

Conclusions

Using the RE_N approach and including global studies conducted in a wide diversity of cropping system, 47 percent of the applied fertilizer N was recovered by the crop (grain and straw) in the year the application occurred. If the cropping systems were in a near steady-state with respect to their N content, the remaining 53 percent of the N applied could be considered lost. Using ^{15}N tracers across a wide diversity of climatic regions and cropping systems, the total amount of N recovered by the first year crop (grain and straw) was estimated at 44 percent. By including the cumulative ^{15}N-fertilizer recovery during the subsequent five growing seasons (6.5 percent) and the amount of ^{15}N recovered in the soil after five growing seasons (15 percent; IAEA, 2003), total ^{15}N fertilizer losses from the different cropping systems would have been 34.5 percent, which is lower than the N losses estimated by the N-balance approach. Possible reasons for the differences in estimates in N fertilizer losses between the two approaches are that the cropping systems under investigations were not at steady state and were still accumulating N. Another major factor that remains is the uncertainty associated with estimating the size of the various N pools and N fluxes.

Independent of the method used to estimate N losses from a cropping system, a further reduction in N fertilizer loss and reactive N will be needed to reduce its negative impact on the well functioning of the biosphere (Boyer and Howarth 2002). The strategy to follow will be (1) to increase direct fertilizer N use by the crop during the year fertilizer N is applied and (2) concurrently to increase the sequestration of fertilizer N not taken up by the crop as soil organic N where it can then serve as a slow release form of N for subsequent crops.

Literature Cited

Baligar, V. C., N. Fageria, and Z. He. 2001. Nutrient use efficiency in plants. *Communication in Soil Science and Plant Analysis* 32:921–950.

Boyer, E. W., and R. W. Howarth. 2002. *The nitrogen cycle at regional to global scales.* Dordrecht: Kluwer Academic Publishers.

Bronson, K. F., J. T. Touchton, R. D. Hauck, and K. R. Kelly. 1991. Nitrogen 15 recovery in winter wheat as affected by application timing and dicyandiamide. *Soil Science Society of America Journal* 55:130–135.

Cassman, K. G., A. Dobermann, and D. Walters. 2002. Agroecosystems, nitrogen-use efficiency, and nitrogen management. *Ambio* 31:132–140.

Carefoot, J. M., and H. H. Janzen. 1997. Effect of straw management, tillage timing and timing of fertilizer nitrogen application on the crop utilization of fertilizer and soil nitrogen in an irrigated cereal rotation. *Soil & Tillage Research* 44:195–210.

Corbeels, M., G. Hofman, and O. van Cleemput. 1998. Residual effect of nitrogen fertilization in a wheat-sunflower cropping sequence on a Vertisol under semi-arid Mediterranean conditions. *European Journal of Agronomy* 9:109–116.

De Datta, S. K., W. N. Obcemea, R.Y. Chen, J.C. Calabio and R.C. Evangelista. 1987.

Effect of water depth on nitrogen use efficiency and nitrogen-15 balance in lowland rice. *Agronomy Journal* 79:210–216.

Dobermann, A., S. Abdulrachman, H. Gines, R. Nagarajan, S. Satawathananont, T. Son, P. Tan, G. Wang, G. Simbahan, M. Adviento, and C. Witt. 2004. Agronomic performance of site specific nutrient management in intensive rice cropping systems in Asia. Pp. 307–336 in *Increasing productivity of intensive rice systems through site specific nutrient management,* edited by A. Dobermann, C. Witt, and D. Dawe. Enfield, New Hampshire and Los Baños, Philippines: Science Publishers Inc., and International Rice Research Institute.

Dobermann A., C. Witt, D. Dawe, S. Abdulrachman, H. C. Gines, R. Nagarajan, S. Satawathananont, T. T. Son, P. S. Tan, G. H. Wang, N. V. Chien, V. T. K. Thoa, C. V. Phung, P. Stalin, P. Muthukrishnan, V. Ravi, M. Babu, S. Chatuporn, J. Sookthongsa, Q. Sun, R. Fu, G. C. Simbahan, and M. A. A. Adviento. 2002. Site-specific nutrient management for intensive rice cropping systems in Asia. *Field Crops Research* 74:37–66.

Eagle, A. J., J .A. Bird, J .E. Hill, W. R. Horwath, and C. van Kessel. 2001. Nitrogen dynamics and fertilizer N use efficiency in rice following straw incorporation and winter flooding. *Agronomy Journal* 93:1346–1354.

FAO (Food and Agriculture Organization of the United Nations). 2004a. *Statistical databases fertilizer consumption.* http://apps.fao.org/page/collections?subset=agriculture.

FAO (Food and Agriculture Organization of the United Nations). 2004b. *Statistical databases cereals production.* http://apps.fao.org/page/collections?subset=agriculture.

Frissel, M. J. 1978. *Cycling of mineral nutrients in agricultural systems.* Amsterdam: Elsevier.

Haefele, S. M., M. C. S. Wopereis, M. K. Ndiaye, S. E. Barro, and M. O. Isselmod. 2003. Internal nutrient efficiencies, fertilizer recovery rates and indigenous nutrient supply of irrigated lowland rice in Sahelian West Africa. *Field Crops Research* 80:19–32.

Hart, P. B. S., D. S. Powlson, P. R. Pulton, A. E. Johnson, and D. S. Jenkinson. 1993. The availability of the nitrogen in the crop residues of winter wheat to subsequent crops. *Journal of Agricultural Science (Cambridge)* 121:355–362.

Howarth, R. W., E. W. Boyer, W. J. Pabich, and J. N. Galloway. 2002. Nitrogen use in the United States from 1961–2000 and potential future trends. *Ambio* 31:88–96.

IAEA (International Atomic Energy Agency). 2003. Management of crop residues for sustainable crop production. IAEA TECHDOC-1354. Vienna, Austria: International Atomic Energy Agency.

Jansson, S. L., and J. Persson. 1982. Mineralization and immobilization of soil nitrogen. Pp. 229–252 in *Nitrogen in agricultural soils,* edited by F. J. Stevenson. Madison, Wisconsin: American Society of Agronomy.

Janzen, H. H., K. A. Beauchemin, Y. Bruinsma, C. A. Campbell, R. L. Desjardins, B. H. Ellert and E. G. Smith. 2003. The fate of nitrogen in agroecosystems: An illustration using Canadian estimates. *Nutrient Cycling in Agroecosystems* 67:85–102

Kumar, K., and K. M. Goh. 2002. Recovery of ^{15}N-labelled fertilizer applied to winter wheat and perennial ryegrass crops and residual ^{15}N recovery by succeeding wheat crops under different crop residue management practices. *Nutrient Cycling in Agroecosystems* 62:123–130.

Ladd, J. N., and M. Amato. 1986. The fate of nitrogen from legume and fertilizer sources in soils successively cropped with wheat under field conditions. *Soil Biology and Biochemistry* 18:417–425.

Oenema, O., and S. Pietrzak. 2002. Nutrient management in food production: Achieving agronomic and environmental targets. *Ambio* 31:159–168.

Panda, M., A. Mosier, S. Mohanty, S. Charkravorti, A Chalam, and M. Reddy. 1995. Nitrogen utilization by lowland rice as affected by fertilization with urea and green manure. *Fertilizer Research* 40:215–223.

Phongpan, S., and A. Mosier. 2003. Effect of crop residue management on nitrogen dynamics and balance in a lowland rice cropping system. *Nutrient Cycling in Agroecosystems* 66:223–240.

Pilbeam, C. J. 1996. Effect of climate on the recovery in crop and soil of N-15 labelled fertilizer applied to wheat. *Fertilizer Research* 45:209–215.

Raun, W. R., and G. V. Johnson. 1999. Improving nitrogen use efficiency for cereal production. *Agronomy Journal* 91:357–363.

Roberts, T. L., and H .H. Janzen. 1990. Comparison of direct and indirect methods of measuring fertilizer N uptake in winter wheat. *Canadian Journal of Soil Science* 70:119–124.

Sheldrick, W. F., J. K. Syers, and J. Lingard. 2002. A conceptual model for conducting nutrient audits at the national, regional, and global scales. *Nutrient Cycling in Agroecosystems* 62:61–72.

Smil, V. 1999. Nitrogen in crop production: An account of global flows. *Global Biogeochemical Cycles* 13:647–622.

Smil, V. 2002. Nitrogen and food production: Proteins for human diets. *Ambio* 31:126–131.

Zapata, F., and O. van Cleemput. 1985. Recovery of ^{15}N labelled fertilizer by sugar-beet spring wheat and winter rye-sugar beet cropping sequences. *Fertilizer Research* 8:269–278.

Zhi-hong, C., S. K. De Datta and I. R. P. Fillery. 1984. Nitrogen-15 balance and residual effects of urea-N in wetland rice fields as affected by deep placement techniques. *Soil Science Society of America Journal* 48:203–208.

15

Pathways and Losses of Fertilizer Nitrogen at Different Scales

Keith Goulding

Nitrogen Loss Pathways and Controlling Factors

The main loss pathways for nitrogen (N) are (1) the runoff or erosion of N in particulate soil organic matter or sorbed on clays; (2) leaching, predominantly of nitrate but also of nitrite, ammonium, and soluble organic N (the last is especially important in grassland; Murphy et al. 2000); (3) gaseous emissions of nitrous oxide and dinitrogen from nitrification and denitrification; and (4) ammonia volatilization (Follett and Hatfield 2001). Research into losses of N at the field scale over the past 20 years shows that they are determined by controllable factors, such as N inputs, crop type and rotation, tillage and land drainage, and uncontrollable factors, such as climate and soil type, described in detail by Balasubramanian et al. and Peoples et al. (Chapters 2 and 5, this volume) and by Hatch et al. (2003).

Of the fertilizer N used to produce the food eaten by livestock, no more than 30 percent is transformed into protein (Oenema et al. 2001). The remaining 70 percent or more is excreted. Losses from excreta during housing and storage can be up to 30 percent of the total N, and an additional 50 percent can be lost during and after application to land. Jarvis (2000) showed that, of the 450 kg N ha⁻¹ applied to a dairy farm in the UK, 36 percent was lost over the whole cycle. With the increasing demand from people in developing countries for more animal protein in their diet and little reduction in the demand in developed countries, an assessment of losses from manures is critical to understanding total losses from fertilizers and is included here.

This review of losses at a range of scales first looks at the data available on a regional basis and then considers the problem of scaling up from laboratory or field experiments to give regional and other large-scale estimates, with or without models, and finally discusses losses at different scales.

Table 15.1. Losses of nitrogen (kt) as N_2O + NO from mineral fertilizers and manures applied to crops or grassland and as NH_3 from mineral fertilizers or manures applied to fertilized grasslands, upland crops, and wetland rice, by region, 1995 (IFA/FAO 2001)

Region	$N_2O–N$ + $NO–N$ From crops	$N_2O–N$ + $NO–N$ From grassland	$NH_3–N$ From manure	$NH_3–N$ From mineral fertilizers	Total gaseous N loss
Canada	170	47	86	140	443
United States	483	81	762	802	2128
Central America	137	29	197	223	586
South America	362	78	567	365	1372
North Africa	77	11	24	230	342
Western Africa	212	56	75	23	366
Eastern Africa	109	30	84	17	240
Southern Africa	96	27	43	54	220
OECD Europe	364	141	1246	607	2358
Eastern Europe	105	21	279	136	541
Former Soviet Union	444	173	1068	217	1902
Near East	128	15	118	443	704
South Asia	800	11	1206	2857	4874
East Asia	666	24	1553	4147	6390
Southeast Asia	332	19	323	756	1430
Oceania	138	69	33	133	373
Japan	23	2	95	64	184
World total	4648	836	7759	11,242	24,485

OECD, Organisation of Economic Co-operation and Development

Regional Analysis of Nitrogen Loss

The International Fertilizer Industry Association and the Food and Agriculture Organization (IFA/FAO 2001) have reviewed the information available about losses of ammonia and nitric and nitrous oxides from fertilizers, with a baseline date of 1995. The report discussed the factors controlling losses and measurement techniques and then analyzed those measurements and produced regional and global estimates of losses. The report included world maps showing the size of loss for a range of crops from each location where measurements have been made. The results are summarized in Table 15.1.

Van Drecht et al. (2003) used a new component of the Integrated Model to Assess

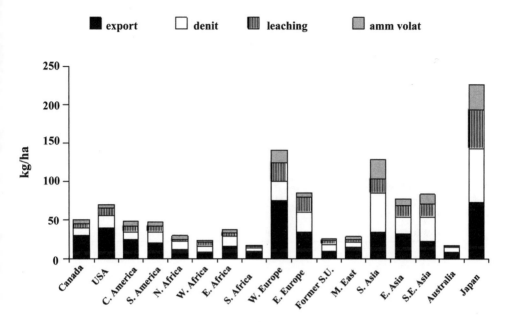

Figure 15.1. Mean regional exports and losses of nitrogen (after Van Drecht et al. 2003).

the Global Environment (IMAGE) to estimate amounts of nitrogen lost by leaching, denitrification, and ammonia volatilization at a spatial resolution of 0.5 by 0.5 degrees for major world regions (Figure 15.1). The results are not directly comparable with those in the IFA/FAO report because the IFA/FAO data are in kilograms per region[-1] and report losses of nitric and nitrous oxides (i.e., potential pollutants), whereas Van Drecht et al. (2003) report losses as kg ha[-1] and total losses by denitrification. A simple conversion from kg ha[-1] to kg region[-1] based on total land areas is not sensible. Comparing the data, however, shows that (1) the most efficient production systems are in the developed world, (2) the largest losses per hectare and in total also tend to be in the developed world, and (3) large losses also occur from flooded rice systems in Asia by ammonia volatilization. The IFA/FAO review, however, shows ammonia volatilization to be much more significant compared with leaching and denitrification than Van Drecht et al.'s model.

The IFA/FAO report and a current search of the CABI and Web of Science databases for publications on nitrogen losses (Table 15.2) show that some regions, such as Europe, North America, Australia, and New Zealand and, recently, China, have conducted a large amount of research; others, such as Africa, most of South and Central America, and the former Soviet Union, have conducted little research. The need for research in these regions is clear.

Table 15.2. Research papers containing the key words *denitrification, ammonia volatilization,* and *nitrate leaching* in CAB International abstracts over the period 1984–2002, by region

Region		Papers on	
	Denitrification	*Ammonia volatilization*	*Nitrate leaching*
Europe	716	185	1126
North America	519	144	561
Latin/Central America	41	32	29
Asia	219	147	159
Africa	61	37	61
Oceania	133	61	151

Europe and North America

In the European Union, the predominant fertilizers used are ammonium nitrate and calcium ammonium nitrate; together they represent more than 40 percent of the total consumption of N-containing fertilizers. Ammonia volatilization from these fertilizers is minimal at 1 to 2 percent of the N applied; leaching and denitrification dominate losses (Figure 15.1). Fertilizer application rates of up to 300 kg N ha^{-1} y^{-1} for cereal crops and 500 kg N ha^{-1} y^{-1} for grass cut for silage and field vegetables are cost-effective in these regions (Goulding 2000).

The environmental impacts can be severe, however. Sanchez (2000) measured the efficiency of use of N (NUE) applied to lettuce grown in the Arizona desert under irrigation. At N and water rates required for maximum yields, about 80 percent of the applied N was not recovered in the aboveground portions of the plant, but losses were not apportioned. The case studies for the U.S. Midwest (Buresh et al., Chapter 10; Murrell, Chapter 11, this volume) and the comparison of agricultural systems in Denmark and The Netherlands (Olesen et al., Chapter 9, this volume) give more detail of such high input systems and show how losses can be controlled.

Australia and New Zealand

Fillery and McInnes (1992) reported that 10 to 40 percent of the N applied to duplex soils in wheat-growing regions of Australia can be lost irrespective of the time of application, with denitrification believed to be the chief cause of loss. For the duplex soils of Western Australia, losses were 50 percent (Palta and Fillery 1993), with circumstantial evidence for volatilization and leaching being the dominant processes.

Weier (1994) reported that up to 50 percent of the urea–N applied to sugarcane was lost; denitrification resulted in losses of 20 percent of applied N on clay soils in sugarcane areas, and ammonia volatilization losses of 10 to 40 percent from urea applied to tropical fruit crops and grassland.

In New Zealand, Cookson et al. (2001) found losses of urea–N applied to perennial ryegrass to be 9 to 23 percent from autumn applications and 19 percent from spring applications because of leaching and denitrification in autumn and ammonia volatilization in spring, a change of dominant loss process with season. When the grassland was ploughed, an additional apparent loss of 11 to 35 percent of the fertilizer N stored in the grassland occurred; that is, total losses were 20 to 58 percent (Williams et al. 2001).

Ledgard et al. (1999) calculated losses on four New Zealand farms of 20 to 204 kg N ha^{-1} by nitrate leaching, 3 to 34 kg N ha^{-1} by denitrification, and 15 to 78 kg N ha^{-1} by volatilization for grazed clover/ryegrass pastures. The total N loss averaged 74 percent of the N applied. By comparison, Cookson et al. (2000) measured leaching losses of only about 8 percent of the 50 kg N ha^{-1} applied to arable land in New Zealand. Kumar and Goh (2002) measured losses of 35 percent of the 120 kg N ha^{-1} applied to winter wheat attributed to leaching and denitrification.

Africa

On-farm experiments in four Sahelian countries between 1995 and 1998 showed average losses of fertilizer N to be 64 percent (Haefele et al. 2003). Apart from these, few data are available. Nutrient depletion is a major problem (Palm et al., Chapter 5, this volume): Stoorvogel et al. (1993) calculated annual N depletion rates for sub-Saharan Africa at 26 kg y^{-1}.

Asia

Pilbeam et al. (1997) using ^{15}N estimated losses from wheat grown in Syria between 1991 and 1995 at greater than 35 percent, mostly by volatilization from the calcareous soil or denitrification from wet soils rich in organic residues.

The IFA/FAO (2001) report has data from China, but much more has emerged since 1995. Roelcke et al. (1996) reported that ammonia volatilization was the major pathway for N loss in the calcareous soils of the Chinese loess plateau, reaching 50 percent of the fertilizer N applied. Cai et al. (2002a) reported that 44 to 48 percent of the urea–N applied to irrigated maize on the North China Plain was lost by volatilization, with denitrification constituting less than 2 percent. Ammonia volatilization accounted for 30 to 40 percent of the N lost from rice and 10 to 50 percent of that applied to maize but only 1 to 20 percent of that applied to wheat growing on a calcareous sandy loam at Fengqiu in the North China Plain (Cai et al. 2002b). Denitrification was not usually a significant pathway of N loss. N losses on eroded sediment from China's Loess Plateau were 40 to 80 kg ha^{-1} y^{-1} (Hamilton and Luk 1993).

Mahmood et al. (2001) reported that up to 42 percent of N applied in crop residues and urea were denitrified during the monsoon season under cotton in Pakistan. More details of this work is presented in the case study in Palm et al. (Chapter 5, this volume).

Latin America

De Koning et al. (1997) reported a depletion of soil N in Ecuadorian agro-ecosystems of about 40 kg ha^{-1} yr^{-1}. Erosion is a major cause of N loss; leaching and denitrification also contribute significantly. Palma et al. (1998) found that 12 percent and 6 percent of urea–N was lost when the fertilizer was surface applied and incorporated, respectively, to "no till" maize in Pampa Humeda, Argentina, and 9 percent and 5 percent, respectively, for conventional tillage.

Scaling

The problem of scaling has been considered by many researchers, mostly in the context of the scaling-up of results using models. It is often assumed that an average value for a loss, measured over a particular time period at small scales, can be simply multiplied up for longer times or larger areas; but multiplying up from measurements over short periods and small scales may not be possible because of the phenomenon of *decoherence*, the unpredictability of measurements at very small scales (Addiscott 1998). As scale increases, processes become more determinate. A good example of this was the observation by Groffman and Tiedje (1989) that predictive relationships between denitrification and environmental factors were easier to establish at landscape than field scale. Models that scale up must be evaluated (Addiscott 1998); however, it becomes more difficult to obtain appropriate data as the scale of use increases.

Scaling-up with models requires some selection of factors to drive the model. Milne et al. (2004) used the Wavelet Theory (a form of geostatistics) to examine the relationship between fluxes of nitrous oxide and their controlling factors. Different factors correlated with fluxes at different scales, and clear evidence of decoherence was found. For leaching, hydrology also becomes much more important as scale increases. Thus, at the watershed scale, N losses can be predicted by simple input/output hydrological models (e.g., Whitehead et al. 1998).

Using small plot experiments and models calibrated and tested at small scales to estimate losses at larger scales must be regarded as a questionable practice unless tests prove the scaling to be appropriate. Pennock at el. (2003) reported the scaling-up from point source measurements using chambers to measure nitrous oxide fluxes during snowmelt at the "township" (92 km^2) scale in Canada. The chambers were carefully placed to reflect the various land uses, with 10 chambers at each site. Scaled fluxes were compared with measurements made from aircraft-based sensors over the two-week snow melt. Emissions differed greatly between sites and could not be explained by soil or climatic factors. Agreement between chambers and aircraft sensors was poor on a day-to-day basis but good when the total for the period was calculated: 58.7 and 47.7 g N_2O-N ha^{-1} for the chamber and aircraft measurements, respectively. Choularton et al. (1995) measured methane effluxes from wetland areas of Scotland using the boundary-layer budget method by collecting air samples with an aircraft upwind and downwind of an area of peat land. The daytime fluxes measured by the aircraft were generally larger than fluxes measured by micrometeorological techniques at the same time and two to four times larger than those measured by cover boxes at the surface.

Losses at Different Scales

Measurements of N loss are made with soil cores or cover boxes at a scale of centimeters, with ^{15}N at the small plot scale of meters, micrometeorology at the field scale (tens to hundreds of meters), and nutrient budgets at the field, farm, national, and regional level of hundreds of meters to kilometers and with aircraft at the national scale of hundreds to thousands of kilometers. For convenience, these will be separated into core and small plot, field and farm, watershed, national and regional, and global scales.

CORE AND SMALL PLOT

As the preceding results show, losses of N measured at small scales are extremely variable in space and time. Coefficients of variation for measurements of leaching with porous-cup tensiometers, a scale of centimeters, can be 90 percent, and annual leaching losses from a field can vary by a factor of 10 even where no N is applied and by a factor of 20 where N is applied because of variations in climate (Goulding et al. 2000). Clearly, short-term experiments at single sites are likely to deliver a wide range of results regarding the pathways and amounts of N lost from fertilizers. If they are to be used in larger-scale budgets, they must be made in sufficient numbers to minimize variations, represent the area adequately, and be continued for at least one year.

FIELD AND FARM

At the farm (as well as regional and national) scale, the calculation of N budgets is a valuable means of indicating the N surplus (the excess of N applied over that in saleable produce) and the potential for, if not the pathway of, loss. The excessive use of N fertilizer has created a large N surplus on some European farms, for example, 320 kg N ha^{-1} in the Netherlands and 170 kg N ha^{-1} in Belgium in the early 1990s (Hatch et al. 2003). For the Broadbalk Experiment at Rothamsted, amounts of N leached are directly proportional to the magnitude of N surplus (Hatch et al. 2003). The link between N surpluses and losses to the environment is clear, and farm-scale research is vital to obtaining appropriate data and finding solutions to N losses.

WATERSHED

Such studies are rare. Shipitalo and Edwards (1998) carried out a 28-year, nine-watershed study on erosion losses in the North Appalachian Experimental Watershed. 92 percent of the erosion occurred during the tillage part of a grass/arable rotation.

NATIONAL AND REGIONAL

Many countries are committed to calculating emissions inventories for ammonia and nitrous oxide (Anon 2003). Emission factors (EFs) are calculated for different soils, climates, crops, and other characteristics. In the case of ammonia, current EFs depend on fertilizer type (anhydrous ammonia > urea > nitrate forms) and cropping. For nitrous oxide, the EF recommended by Mosier et al. (1998) for use by the Intergovernmental Panel on Climate Change (IPCC) is a uniform 1.25 ± 1.0 percent of the N applied. Such an inventory has the benefit of simplicity but reveals no variation with crop or soil, and it implies that only a decrease in N use decreases N losses. Li et al. (2001) compared nitrous oxide emissions from croplands in China, calculated using the process-based DeNitrification–DeComposition (DNDC) model with those made using the IPCC spreadsheet inventory. DNDC and IPCC methods estimated similar total emissions, but geographic patterns were quite different.

GLOBAL

At the global scale, urea constitutes 51 percent of total N use (82.4 Mt total N use; 42.0 Mt urea–N; IFA, 2002). Urea–N is prone to large losses through ammonia volatilization of up to 70 percent (Fillery and McInnes 1992). Bouwman et al. (1997) compiled a global emissions inventory for ammonia (NH_3) showing that about half comes from Asia and about 70 percent is related to food production; the data in Table 15.1 support this view. It should also be noted that about 10 Tg ammonium bicarbonate fertilizer is used in China. Ammonia losses from this are up to twice those from urea (Roelcke et al. 2002). The overall uncertainty in the global emission estimate is 25 percent, whereas the uncertainty in regional emissions is much greater.

Conclusions

Some regions of the world have little data on N losses. Better quantification of losses, especially at the farm scale, linked to the type of input and crop for regions such as Africa, Central and South America, and the former Soviet Union would improve our understanding of the problem, minimize uncertainty in scaling-up, and help toward reducing losses. Loss pathways do not change with scale but can change through the farming year because of climate and management. Current data suggest that about 50 percent of the fertilizer N applied in the world is lost. For European countries in which ammonium nitrate or other nitrate forms dominate fertilizer use, nitrate leaching and denitrification are the main loss pathways. For most of the world, where urea is dominant, ammonia volatilization is the chief loss pathway, especially in warmer climates. Asia is probably responsible for half the ammonia emitted over the world; Asia, Europe, and the United States have the highest emission rates per hectare but also the most efficient farming systems. In many developing and some developed countries, soil N is being depleted by erosion and export in crops.

Literature Cited

Addiscott, T. M. 1998. Modelling concepts and their relation to the scale of the problem. *Nutrient Cycling in Agroecosystems* 50:239–245.

Anon. 2003. *Emission inventory guidebook. Cultures with fertilisers, activities 100101–100105.* Brussels: European Commission.

Bouwman, A. F., D. S. Lee, W. A H. Asman, F. J. Dentener, K. W. Van der Hoek, and J. G. J. Olivier. 1997. A global high-resolution emission inventory for ammonia. *Global Biogeochemical Cycles* 11:561–587.

Cai, G., D. Chen, R. E. White, X. H. Fan, A. Pacholski, Z. L. Zhu, and H. Ding. 2002a. Gaseous nitrogen losses from urea applied to maize on a calcareous fluvo-aquic soil in the North China Plain. *Australian Journal of Soil Research* 40:737–748.

Cai, G. X., D. L. Chen, H. Ding, A. Pacholski, X. H. Fan, and Z. L. Zhu. 2002b. Nitrogen losses from fertilizers applied to maize, wheat and rice in the North China Plain. *Nutrient Cycling in Agroecosystems* 63:187–195.

Choularton, T. W., M. W. Gallagher, K. N. Bower, D. Fowler, M. Zahniser, and A. Kaye. 1995. Trace gas flux measurements at the landscape scale using boundary-layer budgets. *Philosophical Transactions of the Royal Society of London, Series A Mathematical Physical and Engineering Sciences* 351:357–368.

Cookson, W. R., J. S Rowarth, and K. C. Cameron. 2000. The effect of autumn applied N-15-labelled fertilizer on nitrate leaching in a cultivated soil during winter. *Nutrient Cycling in Agroecosystems* 56:99–107.

Cookson, W. R., J. S. Rowarth, and K. C. Cameron. 2001. The fate of autumn-, late winter- and spring-applied nitrogen fertilizer in a perennial ryegrass (*Lolium perenne* L.) seed crop on a silt loam soil in Canterbury, New Zealand. *Agriculture Ecosystems and Environment* 84:67–77.

De Koning, G. H. J., P. J. van de Kop, and L. O. Fresco. 1997. Estimates of sub-national nutrient balances as sustainability indicators for agro-ecosystems in Ecuador. *Agriculture Ecosystems and Environment* 65:127–139.

Fillery, I. R., and K. J. McInnes. 1992. Components of the fertilizer nitrogen-balance for wheat production on duplex soils. *Australian Journal of Experimental Agriculture* 32:887–899.

Follett, R. F., and J. L. Hatfield. 2001. *Nitrogen in the environment: Sources, problems and management.* Amsterdam: Elsevier.

Goulding, K. W. T. 2000. Nitrate leaching from arable and horticultural land. *Soil Use and Management* 16:145–151.

Goulding, K. W. T., P. R. Poulton, C. P. Webster, and M. T. Howe. 2000. Nitrate leaching from the Broadbalk Wheat Experiment, Rothamsted, UK, as influenced by fertiliser and manure inputs and the weather. *Soil Use and Management* 16:244–250.

Groffman, P. M., and J. M. Tiedje. 1989. Denitrification in north temperate forest soils:

Relationships between denitrification and environmental factors at the landscape scale. *Soil Biology and Biochemistry* 21:621–626.

Haefele, S. M., M. C. S. Wopereis, M. K. Ndiaye, S. E. Barro, and M. O. Isselmod. 2003. Internal nutrient efficiencies, fertilizer recovery rates and indigenous nutrient supply of irrigated lowland rice in Sahelian West Africa. *Field Crops Research* 80:19–32.

Hamilton, H., and S. H. Luk. 1993. Nitrogen transfers in a rapidly eroding agroecosystem —loess plateau, China. *Journal of Environmental Quality* 22:133–140.

Hatch, D., K. W. T. Goulding, and D. V. Murphy. 2003. Nitrogen. Pp. 7–27 in *Agriculture, hydrology and water quality*, edited by P. M. Haygarth and S. C. Jarvis. Wallingford: CABI Publishing.

IFA (International Fertilizer Industry Association). 2002. *Fertilizer use by crops*. 5th ed., Rome, Italy: IFA, IFDC, IPI, PPI, FAO. http://www.fertilizer.org/ifa/statistics/crops/fubc5ed.pdf

IFA/FAO (International Fertilizer Industry Association/Food and Agriculture Organization). 2001. *Global estimates of gaseous emissions of NH_3, NO and N_2O from agricultural land*. Rome, Italy: FAO.

Jarvis, S. C. 2000. Progress in studies of nitrate leaching from grassland soils. *Soil Use and Management* 16:152–156.

Kumar, K., and K. M. Goh. 2002. Recovery of ^{15}N-labelled fertilizer applied to winter wheat and perennial ryegrass crops and residual ^{15}N recovery by succeeding wheat crops under different crop residue management practices. *Nutrient Cycling in Agroecosystems* 62:123–130.

Ledgard, S. F., J. W. Penno, and M. S. Sprosen. 1999. Nitrogen inputs and losses from clover/grass pastures grazed by dairy cows, as affected by nitrogen fertilizer application. *Journal of Agricultural Science* 132:215–225.

Li, C. S., Y. H. Zhuang, M. Q. Cao, P. Crill, Z. H. Dai, S. Frolking, B. Moore III, W. Salas, W. Z. Song, and X. K. Wang. 2001. Comparing a process-based agro-ecosystem model to the IPCC methodology for developing a national inventory of N_2O emissions from arable lands in China. *Nutrient Cycling in Agroecosystems* 60:159–175.

Mahmood, T., F. Azam, and K. A. Malik. 2001. Denitrification loss from irrigated croplands in the Faisalabad region—a review of the available data. *Pakistan Journal of Soil Science* 19:41–50.

Milne, A. E., R. M. Lark, T. M. Addiscott, K. W. T. Goulding, C. P. Webster, and S. O'Flaherty. 2004. Wavelet analysis of the scale- and location-dependent correlation of modelled and measured nitrous oxide emissions from soil. *European Journal of Soil Science* (in press).

Mosier, A., C. Kroeze, C. Nevison, O. Oenema, S. Seitzinger, and O. van Cleemput. 1998. Closing the global N_2O budget: Nitrous oxide emissions through the agricultural nitrogen cycle. *Nutrient Cycling in Agroecosystems* 52:225–248.

Murphy, D. V., A. J. Macdonald, E. A. Stockdale, K. W. T. Goulding, S. Fortune, J. L. Gaunt, P. R. Poulton, J. A. Wakefield, C. P. Webster, and W. S. Wilmer. 2000. Soluble organic nitrogen in agricultural soils. *Biology and Fertility of Soils* 30:374–387.

Oenema, O., A. Bannink, S. G. Sommer, and G. L. Velthof. 2001. Gaseous nitrogen emissions from livestock farming systems. Pp. 255–289 in *Nitrogen in the environment: Sources, problems and management*, edited by R. F. Follett and J. L. Hatfield. Amsterdam: Elsevier.

Palma, R. M., M. I. Saubidet, M. Rimolo, and J. Utsumi. 1998. Nitrogen losses by volatilization in a corn crop with two tillage systems in the Argentine Pampa. *Communications in Soil Science and Plant Analysis* 29:2865–2879.

Palta, J. A., and I. R. Fillery. 1993. Nitrogen accumulation and remobilization in wheat of N-15-urea applied to a duplex soil at seeding. *Australian Journal of Experimental Agriculture* 33:233–238.

Pennock, D. J., R. Desjardins, E. Pattey, and J. I. MacPherson. 2003. Multi-scale estimation of N_2O flux from agroecosystems. Final report to Climate Change Funding Initiative in Agriculture, March 2003.

Pilbeam, C. J., A. M. McNeill, H. C. Harris, and R. S. Swift. 1997. Effect of fertilizer rate and form on the recovery of N-15-labelled fertilizer applied to wheat in Syria. *Journal of Agricultural Science* 128:415–424.

Roelcke, M., Y. Han, S. X. Li, and J. Richter. 1996. Laboratory measurements and simulations of ammonia volatilization from urea applied to calcareous Chinese loess soils. *Plant and Soil* 181:123–129.

Roelcke, M., S. X. Li, X. H. Tian, Y. J. Gao, and J. Richter. 2002. In situ comparisons of ammonia volatilization from N fertilizers in Chinese loess soils. *Nutrient Cycling in Agroecosystems* 62:73–88.

Sanchez, C. A. 2000. Response of lettuce to water and nitrogen on sand and the potential for leaching of nitrate-N. *Hortscience* 35:73–77.

Shipitalo, M. J., and W. M. Edwards. 1998. Runoff and erosion control with conservation tillage and reduced-input practices on cropped watersheds. *Soil and Tillage Research* 46:1–12.

Stoorvogel, J., E. Smaling, and B. H. Janssen. 1993. Calculating soil nutrient balances in Africa at different scales .1. Supra-national scale. *Fertilizer Research* 35:227–235.

Van Drecht, G., A. F. Bouwman, J. M. Knoop, A. H. W. Beusen, and C. R. Meinardi. 2003. Global modeling of the fate of nitrogen from point and nonpoint sources in soils, groundwater, and surface water. *Global Biogeochemical Cycles* 17,1115:26.1–26.20.

Weier, K. L. 1994. Nitrogen use and losses in agriculture in subtropical Australia. *Fertilizer Research* 39:245–257.

Whitehead, P. G., E. J. Wilson, and D. Butterfield. 1998. A semi-distributed integrated nitrogen model for multiple source assessment in catchments (INCA): Part 1, model structure and process equations. *Science of the Total Environment* 210/211:547–558.

Williams, P. H., J. S. Rowarth, and R J. Tregurtha. 2001. Uptake and residual value of N-15-labelled fertilizer applied to first and second year grass seed crops in New Zealand. *Journal of Agricultural Science* 137:17–25.

16

Current Nitrogen Inputs to World Regions

Elizabeth W. Boyer, Robert W. Howarth, James N. Galloway, Frank J. Dentener, Cory Cleveland, Gregory P. Asner, Pamela Green, and Charles Vörösmarty

A century ago, natural biological nitrogen (N) fixation was the only major process that converted atmospheric N_2 to reactive, biologically available forms. Since then, human activities have greatly increased reactive N inputs to landscapes. Much of the change in the N cycle stems from (1) the creation of reactive N via the Haber–Bosch process for fertilizers and other industrial applications; (2) cultivation of N-fixing crops; and (3) fossil-fuel burning (Smil 2001). Activities associated with the rising human population have more than doubled the amount of reactive N entering the environment (Galloway et al. 2004) (Table 16.1).

Much of the change in the global N cycle is due to the creation of synthetic fertilizers, which has created reactive N at a rate four times higher than that produced by fossil-fuel combustion (Galloway et al. 2004). The enhanced availability of reactive N provides many benefits, especially increased food production and security (Peoples et al., Chapter 4, this volume), although numerous adverse consequences of increasing N inputs occur, ranging from the effects on ecosystem function to effects on human health (Galloway et al. 2004; Townsend et al. 2003). For example, anthropogenically enhanced N inputs to the landscape have been linked to many environmental concerns, including forest decline (Aber et al. 1995), acidification of lakes and streams (Evans et al. 2001), severe eutrophication of estuaries (NRC 2000), and human respiratory problems induced by exposure to high concentrations of ground level ozone and particulate matter (Townsend et al. 2003). In this chapter, we examine N budgets at regional scales. The geographic units presented in this regional analysis include Africa, Asia, Europe (including the Former Soviet Union [FSU]), Latin America, North America, and Oceania. These units are collections of countries as defined by the Food and Agricultural Organization

Table 16.1. Comparison of reactive nitrogen (Tg N yr^{-1}) from natural and anthropogenic sources in terrestrial lands (after Galloway et al. 2004) in 1860 and 1995

	1860	*1995*
Natural sources	125.4	112
Lightning	5.4	5.4
Biological N fixation	120	107
Anthropogenic sources	15	156
Haber–Bosch	0	100
BNF-cultivation	15	31.5
Fossil fuel combustion	0.3	24.5
Total	141	268

BNF, Biological Nitrogen Fixation.

of the United Nations (FAO 2000). Quantifying the changing N inputs to world regions is critical for mitigating the problems associated with N pollution.

Nitrogen Sources

We quantified inputs of new N to each geographic region of interest utilizing a modification of the N budget method developed by Howarth et al. (1996) for large regions. Our goal was not to quantify the entire distribution of N for each landscape but rather to quantify and sum the new inputs of reactive N to each region from both anthropogenic and natural sources. *New N* refers to reactive N that was either fixed within a region or transported into a region. Anthropogenic N sources include fertilizer, biological N fixation in cultivated cropland, net imports of N in human food and animal feedstuffs (where a negative net import term indicates a region that is a net exporter of food and feed), and atmospheric NO_y-N deposition from fossil-fuel combustion. The natural sources include biological N fixation in natural (noncultivated) land and N fixation by lightning. These represent the total net N inputs per unit area of landscape. Animal waste *(manure)* and human waste *(sewage)* are not considered new N inputs because they are recycled within a region; the N in these wastes originated either from N fertilizer, N fixation in agricultural lands, or N imported in food or feeds. Similarly, deposition of ammonium is not considered a new input because it is largely recycled N volatilized from animal wastes (Boyer et al. 2002).

The budget approach is useful because it allows assessment of the relative importance of the various sources of N to a region and provides a systematic method that enables comparison of the responses among regions over time. All N budget data are presented in units of mass per time (Tg N yr^{-1}; 1 Tg N = 1 million metric tons N). Details of our N budgeting methods are presented in detail in other studies (Boyer et al. 2002; Galloway et al. 2004; Howarth et al. 1996) and thus are described only briefly here. The spatial databases obtained below were assigned to geographic regions needed for this study using political boundaries delineated by the Environmental Systems Research Institute (ESRI 1993).

Nitrogen Inputs from Fertilizers

Globally, the production and application of N fertilizers are the single largest anthropogenic sources of reactive N to landscapes. Whereas synthetic fertilizer inputs were a nonexistent source of new N inputs (0 percent) in 1860, they were the dominant global source of anthropogenic N inputs (63 percent) in the 1990s (Table 16.2). To describe the pattern of N fertilizer use, we used country-level estimates of nitrogenous fertilizer consumption from the FAO (FAOSTAT 2003). The net N input of synthetic N fertilizer in any region represents the difference between creation of N fertilizers in the regions and the net trade (import or export) of fertilizers between regions.

Nitrogen Inputs from Fixation in Cultivated Lands

Reactive N is also introduced to the landscape in significant quantities via biological N fixation (BNF) in cultivated land. Natural biological N fixation accounts for nearly 26 percent of the net anthropogenic N inputs at a global scale in the mid-1990s (Table 16.2). To quantify BNF resulting from human cultivation of crops, we calculated the annual agricultural fixation for 1995 using crop areas and yields reported by the FAO (2002). We multiplied the area planted in leguminous crop species by the rate of N fixation specific to each crop type, assigning rates recommended by Smil (1999, 2001).

Nitrogen Inputs from Fixation in Noncultivated Lands

The vast majority (96 percent) of N inputs from natural sources comes from BNF in natural noncultivated vegetated lands of the world, with the remainder coming from reactive N creation by lightning. BNF in natural systems has decreased by more than 10 percent since 1860 (from 120 Tg N in 1860 to 107 in 1995) as a result of land conversion and removal of natural N-fixing species (Table 16.1). Although total net anthropogenic sources (123 Tg N yr^{-1}) currently outweigh natural inputs from BNF (107 Tg N yr^{-1}) on a global basis, natural BNF remains the dominant input term in Africa, Latin America, and Oceania (Table 16.2).

To estimate natural BNF inputs to each region, we used modeled estimates presented by Cleveland et al. (1999) and modified by Cleveland and Asner (personal communication). Their model is based on estimates of plant N requirement simulated with the TerraFlux biophysical–biogeochemical process model to constrain estimates of BNF in vegetation across biomes of the world. Fixation rates encompassed in the model are

Table 16.2. Input of reactive nitrogen to world regions, mid-1990s (Tg yr^{-1})

Region	Anthropogenic					Natural			Total net inputs
	Fertilizer use	BNF in cultivated lands[1]	Imports in food & feed[2]	Atmospheric deposition[3]	Total net anthropogenic input	BNF in noncultivated lands	Fixed by lightning	Total net natural inputs	
Africa	2.1	1.8	0.5	2.9	7.3	25.9	1.4	27.2	34.5
Asia	44.2	13.7	2.3	3.8	63.9	21.4	1.2	22.6	86.5
Europe & FSU*	12.9	3.9	1.0	2.9	20.7	14.8	0.1	14.9	35.6
Latin America	5.1	5.0	-0.9	1.8	11.1	26.5	1.4	27.9	39.0
North America	12.6	6.0	-2.9	2.7	18.4	11.9	0.2	12.0	30.5
Oceania	0.7	1.1	-0.3	0.3	1.8	6.5	0.2	6.7	8.5
Total	77.6	31.5	-0.3	14.4	123.2	106.9	4.4	111.3	234.5

[1] Biological N fixation.

[2] Net N imports; negative values indicate a net export of N.

[3] Net atmospheric deposition of NO$_y$-N from fossil fuel combustion.

*FSU, former Soviet Union.

based on a synthesis of rates reported in the literature. We used simulations for the mid-1990s, where cultivated areas of the landscape under human control were excluded.

Nitrogen Inputs from Food Transfers

Humans and animals require food and feed, and their nutritional needs are met both through both local agricultural production and importation from other regions. Transfers of agricultural products are not the dominant source of N to continental world regions, but they account for a significant redistribution of N among regions, with some regions receiving net N inputs (Africa, Asia, Europe, and the Former Soviet Union) and some regions being net exporters of N (Latin America, North America, and Oceania). This highlights the disconnection between the sites of food production and consumption and indicates the importance of agricultural trade to the redistribution of N (Table 16.2). We estimated annual net N exports in food and feed for 1995 using import and export data from the FAO agricultural trade databases (FAOSTAT 2003). At a continental scale, we tabulated imports and exports for each major crop type provided by the FAO and their N contents (Bouwman and Booij 1998; Lander and Moffitt 1996). We disaggregated the continental data to the scale of our regions of interest based on their fraction of area within each continent.

The net import of N in food and feed reflects a mass balance of needs versus production and inherently includes food production (grains, vegetables, meat, milk, and eggs) and waste (human septic and sewage and animal manure). For example, the N in animal products can be calculated as the difference between animal feed consumption (N intake in crops) and animal excretion (waste production). We obtained data on N available in waste production (as manure) from Sheldrick et al. (2003), based on FAO animal inventories. We assumed that net N import in food and feed is equal to the difference between N demands for human and animal populations in each region and N produced to satisfy those needs in crop and animal production in each region (Howarth et al. 1996, Boyer et al. 2002). Thus, the "net import in food and feed = human consumption + animal consumption – crop production for animal consumption – crop production for human consumption – animal production for human consumption." Cases where the balances are negative, with crop and animal production exceeding human and animal demands, indicate a net export of N in food and feed.

Nitrogen Inputs from Atmospheric Deposition

The N deposition associated with industrial, automotive, and biogenic N emissions provides significant N input at the regional scale. On a global basis, net inputs from atmospheric N deposition account for about 12 percent of the total anthropogenic inputs to continental world regions (Table 16.2). We consider atmospheric N deposition inputs via oxidized forms (NO$_y$), which come largely from the combustion of fossil fuels (Howarth et al. 1996; Prospero et al. 1996). Globally, the release of reactive N to the environment from fossil-fuel combustion as NO$_x$ is about one quarter the rate of N inputs from the use of inorganic fertilizer (Galloway et al. 2004). We obtained modeled estimates of total (wet + dry) atmospheric deposition of NO$_y$-N from fossil-fuel combustion for 1993 from the global chemistry transport model (TM3) of the University of Utrecht (Lelieveld and Dentener 2000). Note that these data reflect NO$_y$-N deposition as a result of anthropogenically induced fossil-fuel burning, which is a large fraction of NO$_y$-N deposition (Galloway et al. in press). The TM3 model, providing simulations on a 5 by 3.75-degree grid, has been widely used and validated extensively for N species (e.g., Holland et al. 1999).

To avoid double accounting of N in our calculation of new, net atmospheric N inputs, we excluded all N that is both emitted and redeposited within our regional boundary. By assuming the volatilization and deposition cycle of reduced (e.g., NH$_x$) and organic N forms is complete over the cycle of the large region, these N products do not represent new inputs to regions in our N budgeting procedure (Howarth et al. 1996). For example, about 90 percent of NH$_x$ in the atmosphere comes from agricultural sources (Dentener and Crutzen 1994), including animal wastes (manure) and fertilizers. NH$_x$ is short-lived in the atmosphere, with residence times ranging from hours to weeks (Fangmeir et al. 1994) and typically re-deposits within the same region from which it was emitted (Prospero et al. 1996).

Nitrogen Inputs from Fixation by Lightning

Natural lightning formation provides sufficient energy to convert atmospheric N$_2$ to reactive N (Vitousek et al. 1997); however, this is a relatively small source of N in continental world regions (Table 16.2). Lightning accounts for only about 2 percent of the global total net N inputs, and inputs are higher in tropical regions and other regions characterized by high convective thunderstorm activity (Galloway et al. 2004); lightning accounts for roughly 4 percent of total net N inputs in Africa and Latin America. We obtained modeled estimates of total N fixation via lightning for the early 1990s, linked to convection estimates derived from the global chemistry transport model (TM3) of the University of Utrecht (Lelieveld and Dentener 2000) and based on the parameterization of Price and Rind (1992).

Variation of Nitrogen Inputs among World Regions

In 1860, anthropogenic N creation was of only minor importance relative to natural sources. Since then, N fixation in natural ecosystems has decreased by 10 percent, whereas creation by anthropogenic sources has increased more than 10-fold (Galloway et al. 2004, Table 16.1). On a global basis, reactive N inputs to continental landscapes

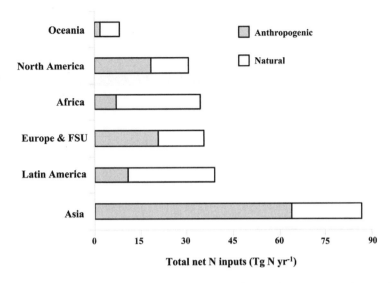

Figure 16.1. Net reactive N inputs to world regions from anthropogenic and natural sources. Anthropogenic sources include N fertilizer use, N fixation in cultivated lands, net N imports in food and feed, and atmospheric N deposition from fossil fuel combustion. The natural sources include biological N fixation in noncultivated vegetated lands and N fixation by lightning. FSU, Former Soviet Union.

from human activities (123 Tg N yr^{-1}) now outweigh N contributions from all natural processes combined (111 Tg N yr^{-1}).

Our N budgets establish total net N inputs to each world region (Table 16.2) and highlight the unequal distribution of new reactive N inputs to the global landscape (Figure 16.1). Natural sources dominate the N budgets in Africa (79 percent), Oceania (79 percent), and Latin America (72 percent), and these large inputs are dominated by natural biological N$_2$ fixation in natural ecosystems. In contrast, anthropogenic N sources dominate the overall N budgets in Asia (74 percent), North America (61 percent), and Europe/FSU (59 percent). Acceleration of the N cycle is affected most significantly in regions of Asia (total inputs = 86 Tg N yr^{-1}). As the region with the highest population and the most intensive and extensive agricultural practices, it also receives the highest N deposition rates globally. Unlike the United States and Europe, which have stabilized rates of population growth, East Asia continues to see rapid increases in population, agriculture, and industrial activity and will continue to play a major role in the global N budget in the future.

Overall, anthropogenic activities related to food production, including N inputs from fertilizers, fixation in cultivated crop lands, and net imports in food and feed have completely altered the global N cycle (Table 16.2). Although the magnitude of N inputs varies

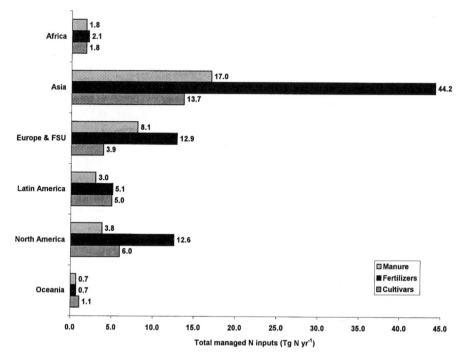

Figure 16.2. Managed N inputs to agricultural lands in world regions from manure applications (available from livestock excreta), from fertilizer use (referring to use of synthetic nitrogenous fertilizers), and in cultivated crop lands (from biological N fixation in legumes, forage, rice, and sugar cane). FSU, Former Soviet Union.

widely by region associated with population and industrial development, synthetic fertilizers are the largest single input of N to most regions. Considering their N contributions relative to the total net new N inputs to each region, the use of fertilizers is largely significant in Asia (51 percent), North America (41 percent), and Europe/FSU (36 percent).

We did not explicitly include inputs of manure in our calculations of total N inputs to each region because manure N is recycled N within a region rather than a newly fixed source. Thus, manure N is accounted for inherently in the term describing net N imports in food and feed (Howarth et al. 1996). To maximize food production, N inputs to agricultural lands are managed and deliberate and come from both recycled sources (from manures, compost, crop residues, or other organic materials) and from newly created N inputs (from mineral fertilizers and fixation in cultivated leguminous crop lands). The relative importance of recycled versus net new inputs to agricultural lands also varies between regions (Figure 16.2). Worldwide, synthetic fertilizers currently account for 54 percent of the managed N inputs to agricultural lands, although contributions from cultivars (22 percent) and manure (24 percent) are also significant. The use of organic manures and cultivars to provide N inputs still outweighs contributions from synthetic fertilizer N use in Latin America, Africa, and Oceania.

Globally, fertilizer use is currently the dominant source of new N inputs to the landscape and is projected to increase significantly in the coming decades, especially in developing regions (FAO 2002; Wood et al., Chapter 18, this volume). Greater N inputs to a region result in a greater potential for N losses (Goulding, Chapter 15, this volume). For example, there is a direct relationship between net N inputs to the landscape and N losses via riverine fluxes (Howarth et al. 1996). The adverse consequences associated with N losses underscore the need to explore strategies that minimize N losses from agricultural lands and maximize N use efficiency. Such strategies will help minimize the adverse effects of adding excess N to the environment while increasing food production and security for people everywhere.

Acknowledgments

We thank the following for their helpful discussions, reviews, and suggestions, which substantially improved the manuscript: Phil Chalk, John Freney, Arvin Mosier, Daniel Mugendi, Keith Syers, and Stanley Wood.

Literature Cited

Aber, J. D., C. L. Goodale, S. V. Olinger, M. L. Smith, A. H. Magill, R .A. Martin, R. A. Hallett, and J. L. Stoddard. 2003. Is nitrogen deposition altering the nitrogen status of northeastern forests? *Bioscience* 53:375–389.

Aber, J. D., A. Magill, S. G. McNulty, R. D. Boone, K. J. Nadelhoffer, M. Downs, and R. Hallett. 1995. Forest biogeochemistry and primary production altered by nitrogen saturation. *Water, Air and Soil Pollution* 85:1665–1670.

Bouwman, A. F., and Booij, H. 1998. Global use and trade of feedstuffs and consequences for the nitrogen cycle. *Nutrient Cycling in Agroecosystems* 52:261–267.

Boyer, E. W., C. L. Goodale, N. A. Jaworski, and R. W. Howarth. 2002. Anthropogenic nitrogen sources and relationships to riverine nitrogen export in the northeastern USA. *Biogeochemistry* 57–58:137–169.

Cleveland, C. C., A. R. Townsend, D. S. Schimel, H. Fisher, R. W. Howarth, L. O. Hedin, S. S. Perakis, E. F. Latty, J. C. Von Fischer, A. Elseroad, and M. F. Wasson. 1999. Global patterns of terrestrial biological nitrogen (N_2) fixation in natural ecosystems. *Global Biogeochemical Cycles* 13:623–645.

Dentener, F. J., and P. J. Crutzen. 1994. A three-dimensional model of the global ammonia cycle. *Journal of Atmospheric Chemistry* 19:331–369.

ESRI (Environmental Systems Research Institute). 1993. *Digital chart of the world*. Redlands, California: ESRI.

Evans, C. D., J. M. Cullen, C. Alewell, J. Kopácek, A. Marchetto, F. Moldan, A. Prechtel, M. Rogora, J. Vesel, and R. F. Wright. 2001. Recovery from acidification in European surface waters. Pp. 283–297 in *Hydrology and earth system sciences*, edited by W. Geller, H. Klapper, and W. Salomons. Berlin: Springer-Verlag.

Fangmeier, A., A. Hadwiger-Fangmeier, L. J. M. Van der Eerden, and H. -J. Jäger. 1994. Effects of atmospheric ammonia on vegetation: A review. *Environmental Pollution* 86:43–82.

FAO (Food and Agricultural Organization of the United Nations). 2000. *Fertilizer requirements in 2015 and 2030.* ISBN 92-5-104450-3, 29 pp. Rome: FAO.

FAO (Food and Agricultural Organization of the United Nations). 2002. *World agriculture: Towards 2015/2040, summary report.* ISBN 92-5-104761-8, 97 pp. Rome: FAO.

FAOSTAT (Food and Agriculture Organization of the United Nations). 2003. FAO-STAT Agriculture Data. [online] URL: http://apps.fao.org/.

Galloway, J. N., F. J. Dentener, D. G. Capone, E. W. Boyer, R. W. Howarth, S. P. Seitzinger, G. P. Asner, C. Cleveland, P. Green, E. Holland, D. M. Karl, A. F. Michaels, J. H. Porter, A. Townsend and C. Vörösmarty. 2004. Nitrogen cycles: Past, present and future. Biogeochemistry (in press).

Holland, E. A., F. J. Dentener, B. H. Braswell, and J. M. Sulzman. 1999. Contemporary and pre-industrial global reactive nitrogen budgets. *Biogeochemistry* 46:7–43.

Howarth, R. W., G. Billen, D. Swaney, A. Townsend, N. Jaworski, K. Lajtha, J. A. Downing, R. Elmgren, N. Caraco, T. Jordan, F. Berendse, J. Freney, V. Kudeyarov, P. Murdoch, Z. Z. Liang. 1996. Regional nitrogen budgets and riverine N & P fluxes for the drainages to the North Atlantic Ocean: Natural and human influences. *Biogeochemistry* 35:75–139.

Lander, C. H., and D. Moffitt. 1996. *Nutrient use in cropland agriculture (Commercial fertilizers and manure): Nitrogen and phosphorus.* Working Paper 14, RCAIII, NRCS, United States Department of Agriculture.

Lelieveld, J., and F. Dentener. 2000. What controls tropospheric ozone? *Journal Geophysical Research* 105:3531–3551.

NRC (National Research Council). 2000. *Clean coastal waters: Understanding and reducing the effects of nutrient pollution.* Washington, DC.: National Academy Press.

Price, C., and D. Rind. 1992. A simple lightning parameterization for calculating global lightning distributions. *Journal Geophysical Research* 97:9919–9933.

Prospero, J. M., K. Barrett, T. Church, F. Dentener, R. A. Duce, J. N. Galloway, H. Levy, J. Moody, and P. Quinn. 1996. Atmospheric deposition of nutrients to the North Atlantic Basin. *Biogeochemistry* 35:27–73.

Sheldrick, W., J. K. Syers, and J. Lingard. 2003. Contributions of livestock excreta to nutrient balances. *Nutrient Cycling in Agroecosystems* 66:119–131.

Smil, V. 1999. Nitrogen in crop production: An account of global flows. *Global Biogeochemical Cycles* 13:647–662.

Smil, V. 2001. *Enriching the earth.* Cambridge, Massachusetts: The MIT Press.

Townsend, A. R., R. W. Howarth, F. A. Bazzaz, M. S. Booth, C. C. Cleveland, S. K. Collinge, A. P. Dobson, P. R. Epstein, E. A. Holland, D. R. Keeney, M. A. Mallin, C. A. Rogers, P. Wayne, and A. H. Wolfe. 2003. Human health effects of a changing global nitrogen cycle. *Frontiers in Ecology and the Environment* 1:240–246.

Vitousek, P. M., J. D. Aber, R. W. Howarth, G. E. Likens, P. A. Matson, D. W. Schindler, W. H. Schlesinger, and D. G. Tilman. 1997. Human alteration of the global nitrogen cycle: Sources and consequences. *Ecological Applications* 7:737–750.

PART VI
Challenges

17

Challenges and Opportunities for the Fertilizer Industry

Amit H. Roy and Lawrence L. Hammond

The principal technology used to produce nitrogen (N) fertilizer today is traced to the Haber–Bosch synthesis of ammonia. The first ammonia plant using this technology began operating in 1913, but inorganic N fertilizer use did not begin to expand dramatically until after World War II. Smil (1999) cites growth of N fertilizer use in the United States where less than 50 percent of U.S. cornfields were fertilized with inorganic N in 1950, but today more than 99 percent are fertilized. The growth is even more dramatic in China, where less than 2 percent of applied N was from inorganic sources in 1950, compared with 75 percent today. The use of N fertilizer in sub-Saharan Africa (SSA) is low today (<1 percent of the world total), but we do not know what the situation will be in the distant future. The dominant N source may change as the need for higher efficiency increases and environmental concerns exert greater pressure.

Nitrogen Fertilizers

Global fertilizer demand, particularly for nitrogenous fertilizers, has been directly related to the demand for food and fiber for the increasing world population, expected to peak at about 8.9 billion by 2050. Synthetic ammonia (NH_3), the principal source of all nitrogen fertilizers, provided only half of the world's inorganic N in 1931; by 1950 that share was almost 80 percent and by 1962, more than 90 percent. During the late 1990s, Haber–Bosch synthesis supplied more than 99 percent of fixed inorganic nitrogen, with the remainder primarily from Chilean nitrate and by-product ammonia from coke ovens.

Nitrogen fertilizers can be classified into four groups depending on their chemical form: ammonium fertilizers, nitrate fertilizers, combined ammonium and nitrate fertilizers, and amide fertilizers. Detailed information regarding characterization and the production technologies for these and other fertilizers is available in the International Fertilizer Development Center/United Nations Industrial Development Organization (IFDC/UNIDO) *Fertilizer Manual*.

Ammonium Fertilizers

Anhydrous ammonia is still the lowest priced N fertilizer because of its high N concentration (82 percent N), but it has not been adopted to any significant degree outside the United States because of safety and environmental concerns. Although ammonium sulfate was once the leading form of N fertilizer, it now supplies a relatively small percentage of the world total because of more rapid growth in the use of urea, ammonium nitrate, and urea–ammonium nitrate (UAN) solutions and anhydrous ammonia. Ammonium sulfate ($[NH_4]_2SO_4$) contains 21 percent N and is available as a by-product from the steel industry and from some metallurgical and chemical processes. One significant source is by-product from the production of caprolactam. Ammonium chloride (NH_4Cl) has been used as a straight N fertilizer or in other grades of compound fertilizers in combination with urea or ammonium sulfate. Containing 26 percent N, its principal raw materials are common salt ($NaCl$) and anhydrous ammonia for the dual-salt process or anhydrous ammonia and hydrochloric acid (HCl) for the direct-neutralization method.

Nitrate Fertilizers

Before the availability of synthetic ammonia, sodium nitrate ($NaNO_3$) of natural origin, primarily from Chile, was the primary raw material for N fertilizer in many countries. Containing 16 percent N and about 27 percent Na, it is a water-soluble fertilizer source used principally for cotton, tobacco, and some vegetable crops. Calcium nitrate, $Ca(NO_3)_2$, contains 16 percent N and is extremely hygroscopic. It is produced primarily in Europe through either neutralization of nitric acid by ground limestone or use of calcium nitrate tetrahydrate by-product separated from nitrophosphate processes.

Ammonium Nitrate Fertilizers

Ammonium nitrate (NH_4NO_3) contains 34 percent N and is produced by reacting anhydrous ammonia and nitric acid. A popular form of nitrogen fertilizer in most European countries and somewhat in the United States and Canada, this fertilizer also has some industrial uses, notably for production of explosives. It is applied as a straight fertilizer or as calcium ammonium nitrate (CAN) (21 percent N).

Amide Fertilizers

Urea, $CO(NH_2)_2$, was first identified in 1773, when it was isolated by crystallization from urine. It was first produced synthetically in 1828 from ammonia and cyanuric acid. The present method of synthesizing urea from ammonia and carbon dioxide first became commercial in 1922 in Germany, 1932 in the United States, and 1935 in England. It contains 46 percent N and is the predominant global N fertilizer. Sulfur-coated or polymer-coated ureas are also marketed to provide controlled-release rates of N from the product.

Multinutrient Fertilizers

Diammonium phosphate (DAP), $(NH_4)_2HPO_4$, and monoammonium phosphate (MAP) are the most popular phosphate fertilizers in the world because of their high analysis and good physical properties. They are produced through the reaction between phosphoric acid and ammonia gas. These fertilizers not only have high phosphate contents, but they also provide much of the N used worldwide. DAP has a grade of 18 percent N and 46 percent P_2O_5; MAP has variable grades of about 11 percent N and 52 percent P_2O_5.

Controlled-release Fertilizers

The two main types of manufactured controlled-release fertilizers (CRFs) are coated fertilizers and slowly soluble urea–aldehyde reaction products (Landels 2003). Commercial urea–aldehyde reaction products include urea–formaldehyde (UF), isobutylidene diurea (IBDU), and crotonylidene diurea (CDU). Coated fertilizers mainly consist of sulfur-coated urea (SCU), polymer- and sulfur-coated urea (P/SCU), and polymer-coated (including resin-coated) fertilizers (PCFs). Today, all SCU produced in North America is P/SCU with a typical grade of 42 percent N and 5 percent S. In 2001 about 30,000 tons of CRFs were consumed in the United States for agricultural crops, whereas 486,000 tons were consumed in nonagricultural markets. Significant quantities are also consumed in Japan, Europe, and Israel.

Nitrogen Demand

Before World War II, global N fertilizer application of three million tons (Mt) to agricultural soils was insignificant. Inorganic N made a significant difference in only a few European countries, Japan, and Egypt. The Netherlands was the most intensive European user of N fertilizers before World War II. Dutch application averaged 50 to 60 kg N ha^{-1}, compared with 20 to 25 kg N ha^{-1} in Germany and less than 3 kg N ha^{-1} in the United States.

Global consumption of N fertilizers in 1949/1950 was about 3.6 Mt N, rose to about 9.2 Mt N in 1960, and more than tripled to 31.7 Mt N by 1970. Despite higher world energy prices, consumption doubled during the 1970s and amounted to 60.7 Mt by 1980. In 1988 consumption reached 80 Mt N. Most of this increase was due to the rapid adoption of new high-yielding varieties of wheat and rice that were more responsive to higher doses of N fertilizers.

After 1988, global use of N fertilizers dropped because of the precipitous decline in demand in the former Soviet Union, post-communist European economies, and most countries of the European Union. Global consumption of N fertilizers fell below 73 Mt N in 1993 and 1994. During subsequent years, N fertilizer consumption slowly increased to about 81.9 Mt N in 2001/2002. Further steady demand should increase the annual consumption to more than 90 Mt N by 2008 (Prud'homme 2003).

Global and Regional Supply/Demand Balance

Besides the differences in N applications of developed and developing economies, regional and national application rates show significant departures from both global and continental averages. The key attributable factors to high rates of N applications are high population densities, scarcity of arable land, and a high share of irrigated cropland. These factors explain Egypt's high use of N, whereas the rest of the African continent consumes less than 3 percent of the global supply of nutrients, although it has about 12 percent of the world's population.

SSA most urgently needs increased fertilizer use because an insufficient supply of nutrients results not only in low crop yields but also in the continuing decline of soil fertility. Recent studies of soil nutrient balance concluded that less than 30 percent of N needed by the region's crops is actually replaced by fertilizers. To reverse this low productivity and soil degradation, the region must significantly increase its use of N, which averages less than 10 kg/ha. To achieve crop production goals established at the World Food Summit, fertilizer use in Africa needs to increase 50 percent by 2015. N use would increase from the current level of 1.4 Mt N to about 5.6 Mt N.

Contrasted with SSA, Asia's food production has increased considerably. The rapidly increasing use of N fertilizers—from 18.6 Mt N in 1975 to the current level of 58.0 Mt N—applied to high-yielding rice and wheat crops has accounted for most of that gain. East Asian gains have been particularly impressive, and China's transition from traditional cropping without inorganic fertilizers to the world's largest user of inorganic N is the best illustration of this rapid change.

Intensive recycling of organic wastes and use of green manures remained the mainstay of China's N supply during 1949 through 1969 following the establishment of the communist regime. Data analysis of past N inputs into China's agriculture shows that synthetic fertilizers provided less than 5 percent of nutrient supply during the early 1950s, and the share was still less than 30 percent of the total by 1970. By the 1980s, inorganic N accounted for about 50 percent of all inputs. The recycling of organic matter, biological fixation, and atmospheric deposition contributed less than 9 Mt N in 1996, compared with 23.6 Mt N applied as inorganic fertilizers during the 1996 crop year. These data imply that ammonia-based compounds have recently been supplying more than 75 percent of all nitrogen.

Table 17.1. Ammonia capacity by region (thousands of Mt N)

Region/Country	1987 Quantity	1987 Share (%)	1999 Quantity	1999 Share (%)	2005 Quantity	2005 Share (%)
China	18,675	16.9	30,450	23.6	33,460	24.6
Former Soviet Union	21,725	19.7	19,340	15.0	18,455	13.6
North America	16,390	14.8	18,955	14.7	18,410	13.6
South Asia	8,935	8.1	15,750	12.2	16,705	12.3
Western Europe	15,635	14.1	11,870	9.2	11,255	8.3
Middle East	4,100	3.7	5,950	4.6	7,795	5.7
Central Europe	9,830	8.9	7,560	5.9	6,820	5.0
Indonesia and Japan	5,800	5.3	7,725	6.0	8,340	6.1
Mexico and Caribbean (including Venezuela)	5,705	5.2	6,415	5.0	8,265	6.1
Other Countries	3,700	3.4	4,790	3.7	6,330	4.7
Total	110,495		128,805		135,835	

Production Versus Importation

Decisions to produce or to import N fertilizers are influenced primarily by the availability of local and external sources of low-cost raw materials (natural gas, naphtha, fuel, and coal) and other imports. The development of the N industry occurred in the developed countries of Western Europe, North America, and Japan through the 1960s. During 1970 through 1980, however, new plant construction shifted to the gas-rich countries of the Caribbean and Middle East. Additional plants were built in some large consuming countries such as China, India, Indonesia, and Pakistan. Similarly, many plant closures occurred in Western Europe and Japan. The Western European share decreased from 20 percent in 1980/1981 to 9 percent in 2001/2002. In 1980/1981, the developing countries accounted for 31 percent of N fertilizer production, and by 2001/2002, their share was 58 percent. The main N-producing regions in 2000/2001 were China (26 percent of world production), North America (16 percent), South Asia (16 percent), former Soviet Union (11 percent), Western Europe (9 percent), and the Middle East (7 percent).

World ammonia capacity increased by nearly 14 percent from 1984 to 1996, whereas capacity for urea increased by 45 percent. The increases were due primarily to (1) a desire by some main importing countries to become more self-sufficient and (2) the construction of export-oriented capacity in the Middle East and the former Soviet Union.

In the future, developing countries are expected to continue to account for most of the increases in ammonia (Table 17.1) and urea capacity. The availability of relatively low-cost feedstock (usually natural gas) will be a main determinant as to where this new capacity is installed. In the mid-1990s, the ammonia industry accounted for about 5 percent of the worldwide natural gas consumption. For economic and environmental reasons, natural gas is the preferred feedstock; however, processes for ammonia production use various energy sources. For example, about 50 percent of China's N fertilizer production is currently based on coal. Natural gas is now the most economical feedstock for ammonia production.

Trade, an important component of the world fertilizer N industry, has increased in recent years in terms of the main N products. The percentage of production that is traded internationally varies from 10 percent for ammonia to about 40 percent for ammonium phosphates. Ammonium phosphate trade increased from 4.2 Mt in 1988 to 8.7 Mt in 2001, or approximately 60 percent of the world phosphate trade.

Challenges and Opportunities for the Fertilizer Industry

The factors that influence the challenges and opportunities for the future of the world N fertilizer industry include (1) population densities that determine demand, (2) the availability of land and irrigation that influence production intensity, (3) the efficiency of nutrient utilization that influences the nutrient required to meet production needs and environmental protection requirements, and (4) the availability of local and external sources of low-cost raw materials (natural gas, naphtha, fuel oil, and coal) and other imports to facilitate economic production of N fertilizers. In this chapter, these factors are considered within three distinct categories of countries (developed countries, countries with large reserves of natural gas, and developing countries that lack reserves of natural gas).

First, the most developed countries of the world (i.e., North America, Western Europe, and Japan) have been both the primary producers and consumers of nitrogenous fertilizers; however, recently these countries have become less competitive compared with those having cheaper sources of natural gas. Likewise, consumption is flat and may even decrease in the future because of environmental concerns and current high rates of application. Many production facilities have either closed or consolidated. In Western Europe, for example, this region's share of the production of N fertilizers dropped from 20 percent of the world's total in 1980/1981 to 11 percent in 1997/1998 (IFA Statistics).

The most important challenge for the industry in these regions is to compete with lower-cost producers by addressing the issues of their own countries (i.e., develop products/methods to maintain high productivity with reduced pollution, segment the market to provide specialty products as opposed to commodities, avoid dependence on foreign producers). Another potential challenge that may someday need to be addressed by the N fertilizer industry is related to two studies that have identified the protein that enables some plant roots to exchange nutrients with microbes, a well-known trait of legumes. In the June, 2002, edition of *Nature*, it was proposed that the obvious next step is to extend nitrogen-fixing to non-symbiotic crop plants. Altering non-legume crops to interact with N-fixing bacteria can be a complex process that may or may not be achieved in the long term, but if it is it could have a tremendous impact on the future consumption of conventional N fertilizers.

Opportunities linked to the challenges within this region may require investment in research and technology to replace commodity products with specialty products. Management strategies may also develop to integrate the use of inorganic N with livestock wastes and other organics. Producers may find it even more important to promote and educate end-users regarding management systems using the most efficient technologies rather than the cheapest fertilizers. They must become recognized as stewards of the public well-being (i.e., productivity and environment).

Today, controlled-release N fertilizers are available that increase the efficiency of uptake by the crop and reduce the entry of N into the groundwater and atmosphere; however, their use is limited primarily to high-value crops (e.g., horticultural, turf grass) because their high production cost makes it uneconomical for field crops compared with commodity sources. In the future, environmental pressures may change the relative economics, and products that are considered expensive today may become standard.

Countries with a contrasting set of challenges and opportunities are those with large reserves of natural gas. North America, Western Europe, and Japan, are being replaced by gas-rich countries like the Soviet Union, the Caribbean, and the Near East, plus large consuming countries such as China, India, Indonesia, and Pakistan. Because natural gas accounts for about three quarters of the variable cost of producing ammonia (Polo 2003), the countries with low-cost natural gas have a significant advantage for N fertilizer production compared with countries that depend on higher-cost natural gas. Ammonia and methanol production accounts for only about 5 percent of the world natural gas consumption; however, about 85 percent of the ammonia is used to produce fertilizer. Urea consumes 45 percent of the world ammonia production (Maene 2001).

The change in production patterns can be observed by changes in global ammonia trade (i.e., increased exports of ammonia indicate increasing production relative to local consumption). In 1999 Russia and the Ukraine accounted for almost one third of the exports of ammonia and the Near East for 11 percent. Russia and the Ukraine also exported 25 percent of the world total of urea, and the Near East exported 11 percent of the total (Maene 2001).

A significant challenge for these countries is how to use the natural gas to produce N at a cost that will not be disruptive to international trade and use the N in domestic markets to ensure that crops receive balanced nutrition (N-P-K). Currently, as illustrated in Figure 17.1 regarding India, producing countries are overusing N relative to P and K. The opportunities for these countries include continued expansion of production to meet the global demands, but without price distortion. They must

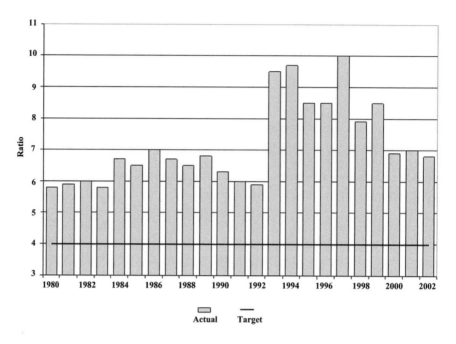

Figure 17.1. N:K ratio in India (Source: T. K. Chanda, Fertilizer Association of India; personal communication).

also learn to be good stewards of the environment. They will now have the opportunity to provide low-cost "commodities" to developing countries that are currently underusing N and need to improve productivity. They also need to educate farmers in developing countries to manage fertilizer use to increase water use efficiency and improve farm income.

One such opportunity for the industry, in this case small-scale entrepreneurs producing urea super-granules (USG) for deep-placement application, is currently being observed in flooded rice-growing regions in Bangladesh and Vietnam more than 20 years after the improved efficiency of this technology was demonstrated in research programs (Mohanty et al. 1999). The practice was not adopted on a large scale sooner primarily because of a lack of a ready supply of USG. Starting in 1996, fabrication of village-level urea briquette compactors based on an IFDC design was initiated in Bangladesh, and this has led to a dramatic increase in the use of USG. Currently, there are 10 manufacturers of the machines in Bangladesh, and more than 1200 of the machines have been produced.

Several hundred field trials have been conducted in South and Southeast Asia by rice agronomists of national and international institutions and networks to evaluate the performance of deep-placed USG (Mohanty et al. 1999). In most cases, the agronomic performance of deep-placed USG was superior to that of two or three split broadcast applications of prilled urea and resulted in an average saving of urea fertilizer of about 33 percent, with an average additional yield of 15 to 20 percent.

Another category of N consumers with a contrasting set of challenges and opportunities include developing countries that lack reserves of natural gas. These countries currently do not use sufficient quantities of N fertilizer to either supply their domestic needs for food grains or to participate in export trade markets. The limiting factor is often the high cost of N relative to the price they can receive for the produce and also the high risk of financial loss as a result of limited or erratic moisture availability. The main growth opportunity for absolute quantities of fertilizer N is in these countries as they intensify production (increases in the future need to come from intensification rather than expansion of cropped area). The challenge for these countries is to allow the farmers to take advantage of the existing production inputs in an economical manner. Constraints to cost reduction are often linked to government policies that increase the cost of importation and distribution. Policies need to be implemented that will encourage market development through the private sector but with government support in the early stages to get them started. Although the importance of using fertilizer is usually recognized even in areas where fertilizer is not used because of high cost, the information needed to manage the fertilizer of optimum productivity is often not available and farmer education is essential. Hunger and poverty are prevalent in these countries and, unless addressed, can pose a significant threat for conflict. Human immunodeficiency virus (HIV) and acquired immunodeficiency syndrome (AIDS) also pose a severe constraint to national development, especially when sufficient nutrition is not provided at attainable prices.

Few short-term opportunities for indigenous N production are available in these countries, but developing countries can show the greatest growth in agricultural production if a climate for adoption of inputs and technologies is fostered. These countries have the most to gain if proper choices are made and the most to lose if not. Policies and mechanisms to promote agricultural production are available. Experiences from countries that have already shown successes with improved crop production in recent years (e.g., Brazil, India, Bangladesh) prove that it can be done, and the lessons from those countries can show the way.

One of the most critical regions in this category is SSA, where the resource base is inherently low and the cost of inputs is excessively high. The highly weathered soils have low content and poor quality organic matter, low levels of N and phosphorus, and low water-retention capacity. Despite the critical need to build soil fertility, these soils also consistently exhibit the lowest rate of fertilizer use on a per hectare basis. High farm-gate prices are primarily the result of high transportation costs because of poor transport infrastructure and the inability to take advantage of economics of large-scale importations. The lack of financial resources and available credit also hinders intensified use of inputs by the farmers.

Because of the low fertilizer consumption in SSA, high-volume fertilizer producers

Table 17.2. Integrated Soil Fertility Management (ISFM) improvement of crop yield and fertilizer profitability in West Africa

	Farmer's Practice		After 4 Years of ISFM	
	Cereal Yield (kg/ha)	VCR Fertilizer [1]	Cereal Yield (kg/ha)	VCR Fertilizer [1]
Maize: bush field	750	— [2]	2750	4
Maize: compound field	3000	— [2]	4600	12
Sorghum	1000	— [2]	1800	8
Cotton	1150	5	2000	8
Irrigated rice	3000	8	5500	12

[1] Value incremental yield/fertilizer cost.
[2] No fertilizer use by farmers.

currently are not interested in lowering inputs costs and probably will not do so until intensification occurs. Potential does exist, however, for an improved inputs market in this region as demonstrated by programs such as the Integrated Soil Fertility Management (ISFM) established in West Africa by IFDC. ISFM technologies integrate the use of inherent soil nutrients, crop residues, compost, and manure with mineral fertilizers to increase productivity while maintaining or enhancing the agricultural resource base. As shown in Table 17.2, farmers using ISFM technologies have improved their situation through increased yields and more responsive soils. Average maize yields in the area are about 1-2 t ha^{-1}; average values for the trials of the participating farmers are between 2.5 and 5.0 t ha^{-1}. By adopting ISFM technologies, farmers are attaining value:cost ratios well above 2. As production and incomes increase over time, the ability of the farmers to purchase inputs will also increase and the potential market for fertilizer producers will improve.

In summary, a critical need in developing countries is to increase fertilizer use, and a main requirement for the future in each of these regional categories is to optimize fertilizer N efficiency use. This issue is important in all production segments, in developed counties to mitigate the effect on the environment and elsewhere because of the unavailability of additional farmland for expansion of cultivated areas. It can be addressed both by modification of N fertilizer sources and by management of the fertilizer in the field.

Subsidy and Nitrogen Fertilizer Use

Fertilizer subsidies have long been a popular option for stimulating fertilizer use when national goals in developing countries have focused primarily on food security and self-sufficiency. Many developing countries have achieved these objectives; however, subsidies entail (1) increased cost to the government, (2) difficulty in administration, (3) resource misallocation, and (4) environmental impact. For example, fertilizer subsidy policies in some countries grossly distort the relative prices of the three primary plant nutrients: N, phosphate, and potash. For example, Indian policies have kept the maximum sales price of N low relative to phosphate and potash, (deregulated in August 1992). As a result, N use has increased sharply compared with phosphate and potash use. The recommended target nutrient ratio ($N:P_2O_5:K_2O$) is reported to be 4:2:1. As shown in Figure 17.1, the $N:K_2O$ ratio surged to almost 10:1 following deregulation in 1992. The $N:P_2O_5:K_2O$ currently is 6.5:2.5:1. This increased level of N use partially contributes to lower yields besides increasing N losses to the environment.

Literature Cited

IFDC/UNIDO (International Fertilizer Development Center/United National Industrial Development Organization). 1998. *Fertilizer manual.* Dordrecht: Kluwer Academic Publishers.

Landels, S. P. 2003. *Global update on slow-release fertilizers.* Paper presented at the 53rd Annual Meeting of the Fertilizer Industry Round Table, October 27–29, 2003, Winston-Salem, North Carolina.

Maene, L. M. 2001. *The challenges and opportunities for the fertilizer industry in a rapidly growing global environment and the important role of the Arab countries in this development.* 4th Petrochemical Conference on Arab Petrochemical Industries Development, May 7–8, 2001, Bahrain.

Mohanty, S. K., U. Singh, V. Balasubramanian, and K. P. Jha. 1999. Nitrogen deep-placement technologies for productivity, profitability, and environmental quality of rainfed lowland rice systems. *Nutrient Cycling in Agroecosystems,* 53:43–57.

Polo, J. R. 2003. *Basic economics of the fertilizer industry.* Paper presented at the IFDC/IFA Phosphate Fertilizer Production Technology Workshop, September 15–19, 2003, Brussels, Belgium.

Prud'homme, M. 2003. *World fertilizer supply and demand outlook.* Paper presented at the Eighth China National Fertilizer Market Symposium, 10–13 November 2003, Yichang City, China.

Smil, V. 1999. *Long-range perspectives on inorganic fertilizers in global agriculture.* Second Travis P. Hignett Memorial Lecture, IFDC-LS-2, IFDC, Muscle Shoals, Alabama.

18

The Role of Nitrogen in Sustaining Food Production and Estimating Future Nitrogen Fertilizer Needs to Meet Food Demand

Stanley Wood, Julio Henao, and Mark Rosegrant

Perspectives on future nitrogen (N) fertilizer use are of interest to many stakeholders, including fertilizer producers and traders who serve commercial farming interests, environmentalists concerned with local and global ecosystem impacts of N fertilizer use, and national and international development specialists concerned with the poverty and food security implications of low yields and declining soil productivity. In this chapter we review available global-scale projections of N fertilizer use and describe two contemporary sets of projections undertaken by the authors.

Changing Structure of Food Demand

Aggregate demand for food is driven by four principal factors: population, income, food prices, and food preferences. Over the past 40 years (1961–2001), the world's population has doubled from 3.1 to 6.1 billion people (growing around 1.74 percent per year), gross domestic product (GDP) per capita, a widely used proxy for average income, has increased from some $10,157 to $29,215 in constant 1995 U.S. dollars, an increase of around 110 percent (World Bank 2003), and average food prices have declined by around 40 percent (IMF 2003). These changes helped spur food consumption by around 260 percent such that, by 2001, even with three billion extra people, per capita food consumption globally had increased by about 30 percent.

The 2.4 percent annual growth in food consumption between 1961 and 2001 was accompanied, however, by a 4.5 percent per year increase in fertilizer N use. So it is clearly important to look beyond aggregate food demand to assess potential future

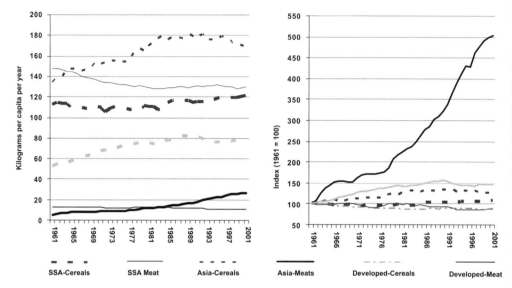

Figure 18.1. Trends in per capita food consumption for selected regions 1961–2001. SSA, sub-Saharan Africa.

change in fertilizer use; it turns out that recognizing and taking account of changing consumer preferences on the structure of food demand is particularly important. Food preferences have and will continue to shift over time with the growing prosperity and education of consumers and amid increasing consumer concerns about issues such as food safety, nutrition, and the environment. Figure 18.1 shows how consumption of two major food categories, cereals and meat, have changed regionally over the past 40 years, sometimes dramatically, as in the case of meat products in Asia. As a consequence of surging demand for meat, coupled with increased economic incentives for confined livestock operations, demand for feed, particularly for quality grains in poultry and pig production, also expands. This process scales up the amount of fertilizer N required to deliver the daily diet. For example, Smil (2002) estimated that in the United States it takes 4.2 kg, 4.2 kg, 10.7 kg, and 31.7 kg of cereal, and hence correspondingly higher N inputs, to produce 1 kg of eggs, chicken, pork, and beef, respectively. For this reason, we rely heavily on the use of food projections by specific food type to gauge more accurately the implications for future fertilizer use.

Existing Projections of Global Nitrogen Fertilizer Use

A review of recent literature on N use projections is summarized in Table 18.1. Clearly, the range of time horizons of interest to analysts is broad, from short-term, fertilizer market–focused projections to 2007/2008 (FAO 2003) to long-term, global-change oriented studies looking out to 2070 (Frink et al. 1999). Interest in 2015 is linked to the Millennium Development Goals and related programs (UN 2002), whereas 2020 and 2030 are time horizons adopted by the International Food Policy Research Institute (IFPRI; Rosegrant et al 2001) and the Food and Agriculture Organization of the United Nations (FAO; Bruinsma 2003), respectively, to focus longer-term efforts on agricultural development linked to hunger- and poverty-reduction strategies. Most studies produced several estimates for their chosen time horizons so as to compare different methods or future scenarios.

Bumb and Baanante (1996) generated projections of fertilizer N to 2000 and 2020 from a baseline of 79.2 Mt per year in 1990, using three approaches. Two were based on assessing the N requirements to meet projected cereal needs in 2020 (Rosegrant et al. 1995): the "nutrient removal approach" and the "cereal production method"), and the third, "effective fertilizer demand," projected N use on the basis of a range of economic, demographic and other factors. These three approaches predicted global fertilizer N use in 2020 to be 203, 146, and 115 Mt, respectively (Table 18.1).

Daberkow et al. (1999) built on crop area and yield projections developed by FAO in support of the Agriculture Toward 2015/2030 study to assess corresponding fertilizer needs. They utilized the *Fertilizer Use by Crop* database (henceforth FUBCD; IFA, IFDC, FAO 1999) to derive crop-specific nutrient application and response rates. Daberkow et al. developed three scenarios: "baseline," "improved nutrient use efficiency" (NUE), and "nitrogen use on cereals," for which projected N fertilizer needs were 100, 88, and 106 Mt in 2015, and 118, 96.2, and 125 Mt in 2030, respectively.

All four Frink et al. (1999) scenarios were based on producing sufficient calories (10,000 calories per person per day) for 10 billion people in 2070. The scenarios were that (1) farm productivity would remain at its 1990 levels with both crop yields and NUE remaining constant, (2) fertilizer would not be applied and production would rely on N deposition from the atmosphere, (3) crop yields would grow at modest rates with improved NUE, and (4) crop yield growth could be maintained at its pre-1990 historic rate with improved NUE. The total fertilizer N requirements under these four scenarios were 284, 0, 192, and 192 Mt, respectively.

For the period 1960 to 1999, Tilman et al. (2001) made several linear regressions between N fertilizer use and time, population, and GDP. Using these regressions, mean values of N fertilizer use were projected as 135 Mt in 2020 and 236 Mt in 2050. The FAO (2003) assessment of near-term fertilizer needs is based on recent market trends including production, consumption, trade, and stocks of N fertilizer at global and regional scales. Global fertilizer N use is projected to grow at around 1.8 percent per year from 2002/2003 to reach 92 Mt in 2007/2008. Galloway et al. (2004) based their N fertilizer projections to 2050 on the Daberkow et al. "baseline" scenario for 2030 and extrapolated an N fertilizer use of 135 Mt in 2050. Cassman et al.

Table 18.1. Global projections of fertilizer nitrogen use

Study	Class G/R	Class F	Base Year	Base N Use	2000	2007/8	2015	2020	2025	2030	2050	2070	Scenario Details
(a) Bumb and Baanante, 1996	R	F	1990	79.2	83.4			115					"Effective demand" approach
	R	F	1990	79.2				203[1]					"Nutrient removal" approach
	R	F	1990	79.2				146[1]					"Cereal production" approach
(b) Daberkow et al. 1999 (adapted for Bruinsma 2003)	R	F	1995/1997	77.8			100	(106)		118			Constant NUE, based on global NPK response
	R	F	1995/1997	77.8			88.0	(91)		96.2			Improved NUE, based on global NPK response
	R	F	1995/1997	77.8			106	(112)		125			Constant NUE, based on global N response
(c) Frink et al. 1999	G		1990	79								284	Productivity stagnation at 1990 conditions
			1990	79								0	Rely on N deposition only
			1990	79				(110)				192	Slow yield growth. NUE = 100%
			1990	79				(110)				192	Recent yield growth trends. NUE = 100%

Study	G/R	F	Base year	Base value				Notes
(d) Tilman et al. 2001	G		2000	87		135	236	Av. of time, population, GDP based projections
(e) FAO 2003	R		2001/2002	81.1	92.4			Short-term trend and outlook based projections
(f) Galloway et al. 2004	R	F[1]	~1995	78		(100)	135	Fertilizer production 145. Based on (b).
Major cereals only (rice, wheat and maize)								
(g) Cassman et al. 2003	G	F	2000[2]	44.4		71.6		Decreased NUE (–15%)
						60.9		Constant NUE (= 40 kg grain / kg N applied)
						53.0		Increased NUE (+15%)

Source: Compiled by authors.

Notes: G/R, only global (G) or includes regional (R) projections (regions not consistent across studies). F, Based on food projections. NUE, nitrogen use efficiency. Parenthesized italics under 2020 indicate linearly interpolated values for purposes of the comparative review. See text and study sources for more detailed descriptions of the studies and scenarios listed.

[1] Based on food projections out to 2030, extrapolated to 2050 using constant growth rate.

[2] Food projections based on 1995 baseline. Fertilizer use and efficiency data calibrated to 2000.

(2003) made projections of N fertilizer needs for cereals ranging from 52 Mt to 72 Mt under various combinations of changes in NUE and harvested area linked to projected food needs in 2025.

Updating and Extending Nitrogen Fertilizer Use Projections

A goal of this chapter is to take a 50-year perspective on N fertilizer needs to mesh with the time frames adopted in other chapters of this volume and to do so at both global and regional scales. Of the available projections, only that of Galloway et al. (2004) covers the required time frame, has region-specific projections, and makes projections based on the demand and supply of agricultural commodities. The Galloway et al. projections, however, are based on the Daberkow et al. study that projects only to 2030, after which Galloway et al. assume a constant N fertilizer growth rate to 2050. Furthermore, the underlying Daberkow et al. study relied on older data sources both for the FUBCD and for regional food projections than were available in 2003. The authors therefore undertook a set of fertilizer N projections to update the available estimates for 2050 and added a medium-term horizon of 2020 to facilitate comparison with existing studies. To complete the suite of time frames, the FAO forecasts to 2007/2008 were adjudged by the authors to be the most authoritative shorter-term projections.

For both 2020 and 2050, two sets of global and regional projections were made. The first used a trend analysis, and the second was based on future food needs. The trend analysis assumed that N fertilizer applications would be higher in areas of significant soil degradation between 2020 and 2050 as part of a broader strategy of soil fertility restoration. Future food needs for the second set of projections were derived using IFPRI's IMPACT model (Rosegrant et al. 2002b).

Trend-Analysis Projections

These projections were made by updating the Bumb and Baanante (1996) "effective demand approach." This involved extrapolating time series of fertilizer N consumption, production, and trade as well as crop area and yield since 1969 using moving average techniques. The crops included were wheat, rice, barley, millet, and maize. The updated application of this approach puts global N fertilizer use at 112 Mt in 2020 compared with 115 Mt of Bumb and Baanante (1996). In extending these projections to 2050, additional factors considered were longer-term population projections and more disaggregated NUE estimates. Bumb and Baanante (1996) originally used one of only two values: one ton of NPK yields 10 tons of cereals in developing countries, and 15 tons in developed. In this assessment, maps of cereal production were overlaid with agroecological zone maps, and NUEs in the range of 10 to 20 tons of cereal per ton of N fertilizer were assigned to each region based on a subjective interpretation of local agroecological and crop management conditions. This included considering whether increased production in any location would likely arise through intensification (e.g., improved seeds and increased fertilizers) or through area expansion.

Given concerns about the food security and poverty implications of long-term nutrient mining, projections beyond 2020 up to 2050 made specific allowance for additional fertilizer application to stabilize and rehabilitate areas of depleted soil fertility as defined by overlaying maps of nutrient depletion. This was seen as a critical strategy where local expansion of food production will be a long-term priority (e.g., poor countries with significant population growth, such as are common in sub-Saharan Africa, but including many hillside and mountain areas in Asia and Latin America).

N use was projected for cereals only, and conversion to total N (for both 2020 and 2050) was made by assuming that cereals will account for 60 percent of total N consumption. No improvement in NUE was considered. The resulting global N fertilizer projection for 2050 was 171 Mt. Given both the conservative assumptions about constant NUE and the goal of soil rehabilitation, this is likely an upper bound on N fertilizer needs. Both the global and regional results of these projections are summarized in Table 18.3. The 2020 results are found in column 4 and the 2050 results in column 7.

Food Production–based Projections

THE IMPACT MODEL AND FOOD PROJECTION RESULTS FOR 2020 AND 2050

The International Model for Policy Analysis of Agricultural Commodities and Trade (IMPACT) represents the global agricultural market for 32 crop and livestock commodities, including all cereals, soybeans, roots and tubers, meats, milk, eggs, oils, oilcakes and meals, sugar and sweeteners, fruits and vegetables, and fish. IMPACT comprises 43 different countries or regions, each with its conditions for supply, demand, and prices for agricultural commodities that are linked through trade, highlighting the interdependence of countries and commodities through global agricultural markets. World agricultural commodity prices are determined annually at levels that clear international markets. Demand is a function of prices, income, and population growth, and growth in crop production in each country is determined by crop prices and the rate of productivity growth. The IMPACT model seeks to minimize the sum of net international trade for each commodity at a world market price that satisfies market-clearing conditions.

IMPACT generates annual projections of crop area, yield, and production; the demand for food, feed, and other uses; prices and trade; and livestock numbers, yield, production, demand, prices, and trade. The base year for the projections used for this analysis is 1997 (using a 3-year average of 1996–1998), and the model incorporates data from a range of sources, including FAO, World Bank, and the United Nations (UN) as well as a system of supply and demand elasticities. IMPACT supports scenario analysis through the adjustment of factors with potentially significant impacts on the future world food situation, including population and income growth, the rate of growth in crop and livestock yields, feed ratios for livestock, investments in agricultural research and irrigation, commodity price policies, and elasticities of supply and demand. IMPACT does not model nutrient input.

The food projections used here are taken from IMPACT's "business as usual" (BAU) scenario with projections to 2050. The BAU scenario simulates the future of food supply and demand if current economic and technical trends continue, for example, in crop area expansion and productivity growth. In the BAU scenario:

- Global population grows at a medium-to-low rate.
- Income levels are medium but increasing.
- Income distribution is moderate but becoming more equitable.
- Investment in agricultural research and technology continues at a moderate rate.
- Irrigation efficiency and water use efficiency improve at a medium rate.
- Irrigated area expands at a medium rate.
- Trade continues to be subject to the existing trade barriers.

IMPACT projections were made for both 2020 and 2050. A regional summary of the production of selected crops for the 1997 base period as well as those projected for 2020 and 2050 are shown in Table 18.2.

The medium levels of agricultural investments and improvements in the efficiency of water use in the BAU scenario lead to a moderate increase in global cereal yields between 1997 and 2050 as follows: wheat 51 percent (58 developing, 42 percent developed); rice 68 percent (70/38); and maize 54 percent (79/42). In the case of "other coarse grains" (including barley, millet, oats, rye, and sorghum), yields are expected to grow by 46 percent (74/40), and soybean yields should increase by 47 percent globally (59/41). IMPACT also assesses changes in harvested area and, hence, in overall production, key factors in assessing the likely consequences for N fertilizer use. Globally, cereal production will increase as follows: wheat 49 percent (67/33), rice 51 percent (54/21), and maize 71 percent (111/39). In the case of coarse grains, production is estimated to grow by 52 percent (103/25). Soybean production increases by 70 percent (110/33).

CONVERTING IMPACT FOOD PROJECTIONS TO NITROGEN FERTILIZER PROJECTIONS

We adopted two N use projection scenarios drawing on the IMPACT food projections for 2020 and 2050, holding constant NUE at 1997 levels and increasing NUE over time. For the first scenario, we adopted the "nitrogen use on cereals" (constant NUE) approach of Daberkow et al. (1999) and first estimated global average N response coefficients for wheat, rice, and maize using version 5 of the FUBCD. Using these response coefficients plus the regional changes in cereal areas and yields projected by IMPACT for 2020 and 2050, we were able to compute implicit levels of cereal N use under constant NUE for those years. For the second scenario, we implemented the Daberkow et

Table 18.2. Past (1997) and projected production (Mt) of selected crops in 2020 and 2050 (IMPACT model "business as usual scenario")

Region	Wheat 1997	Wheat 2020	Wheat 2050	Rice 1997	Rice 2020	Rice 2050	Maize 1997	Maize 2020	Maize 2050	Other coarse grain 1997	Other coarse grain 2020	Other coarse grain 2050	Potatoes 1997	Potatoes 2020	Potatoes 2050	Soybeans 1997	Soybeans 2020	Soybeans 2050
USA	66	85	97	5	6	6	239	288	331	27	32	35	22	24	26	71	88	95
EU	99	104	100	2	2	2	37	39	37	66	69	66	48	46	41	1	2	2
OECD Total	215	255	273	16	17	12	284	338	381	123	142	152	80	80	78	75	93	100
Eastern Europe	31	38	42	0	0	0	29	37	47	19	26	31	33	34	35	0	1	1
FSU	68	88	102	1	1	1	7	10	12	48	53	53	69	65	58	0	0	0
Latin America	23	38	47	13	21	24	74	114	159	14	23	31	15	22	27	45	68	92
SSA	5	8	13	7	16	36	36	60	95	33	56	91	4	6	9	1	1	2
WANA	51	74	98	5	8	9	10	12	14	20	28	34	15	24	29	0	0	0
S Asia	89	126	154	111	172	199	13	18	23	22	28	33	24	44	80	6	11	18
SE Asia	0	0	0	91	127	151	20	29	39	0	0	1	2	3	4	2	3	4
E Asia	115	141	159	139	155	146	123	183	247	14	18	23	58	75	90	15	21	29
Asia Total	204	267	314	341	454	496	157	231	309	36	47	56	84	122	174	23	35	51
ROW	0	0	0	0	0	0	1	1	1	0	0	0	4	6	8	0	0	0
World	597	768	889	384	516	579	596	802	1019	294	374	448	304	359	419	145	198	246
Developed	317	385	421	17	18	13	329	397	457	190	221	238	183	182	173	76	94	101
Developing	280	383	468	367	498	565	267	404	563	104	153	211	121	178	246	69	104	145

Data Source: IFPRI Impact Model 2003.

USA, United States of America; EU, European Union; FSU, Former Soviet Union; OECD Countries, Australia, Canada, EU, Iceland, Japan, New Zealand, Norway, and USA; SSA, Sub-Saharan Africa; WANA, West Asia and North Africa; ROW, rest of world.

al. "improved nutrient use efficiency" approach, but instead of using total fertilizer (NPK) response relationships, we used the same N fertilizer response coefficient derived for our first scenario. This yielded a new set of 2020 and 2050 projections for N fertilizer use on cereals that embedded region-specific NUE improvements. The NUE increases averaged 17 percent in 2020 and 30 percent in 2050 globally, relative to the 1997 NUE base.

To scale projected N use for cereals to cover all crops, we first used the FUBCD to compute the quantity of N fertilizer being used for non-cereal crops in each IMPACT region in 1997 (some 53 percent of total N fertilizer globally). The approach was then to scale up the 1997 N fertilizer usage on the basis of the IMPACT model's projected increase in the production of non-cereals by 2020 and 2050. This was complicated by incompatibilities between the list of non-cereal crops in FUBCD and in the IMPACT model. To circumvent this problem we first computed the total dollar value of production of IMPACT's non-cereal crops in 1997, some 29 commodities. We then divided the quantity of N fertilizer applied to non-cereals by this total value of non-cereal production to derive a 1997 ratio of quantity N fertilizer applied per dollar value of non-cereals. In 2020 and 2050, we applied the same set of crop prices used for 1997 to derive total value of production of non-cereals for those years and converted those dollar values to equivalent amounts of N fertilizer using the 1997 "applied N per dollar of non-cereal crop" ratio. In the case of the increased NUE scenario, we scaled down the 2020 and 2050 non-cereal crop N projections using the same region-specific average NUE gains as computed for cereals.

On the basis of this approach, the projected N fertilizer consumption in 2020 and 2050 under the constant NUE assumption will be 112 Mt and 121 Mt, respectively. Assuming improved NUE, the corresponding quantities are 96 Mt and 107 Mt, respectively. Both the global and regional results for each of these scenarios are summarized in Table 18.3 (columns 5, 6, 8, and 9).

Review of Global and Regional Nitrogen Fertilizer Projections

Studies designed primarily to support analysis of the environmental consequences of N fertilizer use adopt longer time horizons (Frink et al. 2070; Galloway et al. 2050; Tilman et al. 2050). They are also more likely to assess only global projections and not use food projections as a basis (Frink et al. 2070; Tilman et al. 2050). Galloway et al. have both regional breakouts and draw from global food projections but extrapolate regional food projection results from 2030 out to 2050 assuming a constant growth rate. The food projection–based studies are split between those that source the FAO 2015/30 food projections (Daberkow et al.; Galloway et al.) and those that use various rounds of IMPACT model results (Bumb and Baanante, Cassman et al., and the authors). The three generations of IMPACT results were benchmarked on 1990

(Rosegrant et al. 1995) used in Bumb and Baanante, 1995 (Rosegrant et al 2002a) used in Cassman et al., and 1997 (Rosegrant et al. 2002b) used by the authors.

Food-based projections often make a serious attempt only to project cereals (usually defined as maize, rice, and wheat) and convert projections to include all agricultural output based on simple proportions (Bumb and Baanante, the authors' trend analysis). Cassman et al. present only results for cereals. Only Daberkow et al. and the authors' food projection–based analyses attempted to account for the changing structure of demand for cereals as well as other crops over time as revealed by detailed food-projection model results.

To facilitate comparison with existing projections from Table 18.1 we interpolated to 2020 those projections with longer time frames (indicated by the italicized figures in parentheses in the 2020 column). We see both a fair degree of spread and some clustering of projected (and interpolated) estimates of total N fertilizer needs in 2020. These range from 91 Mt (Daberkow, et al.'s "improved nutrient efficiency scenario") to 203 Mt (Bumb and Baanante's "nutrient removal approach"). Not surprisingly, projections based on similar approaches yield similar results. The authors' updated Bumb and Baanante trend analysis estimated 112 Mt compared with the original 115 Mt, suggesting little net impact of updating baseline values from 1990 to 1997 through 2000. Similarly, the Daberkow et al. (interpolated) assessment of 91 Mt of fertilizer N in 2020 under improved total (NPK) fertilizer efficiency is close to our own 96 Mt with improved NUE using more up-to-date base data and a different source of food projections. Evidence has been found that food-based projection approaches are, over time, leading to lower estimates of projected fertilizer use because population as well as GDP growth, and hence future food demand, continue to moderate relative to prior estimates in most parts of the world (but still remain a formidable food security challenge).

Three of the studies considered more than just N fertilizer amounts and considered sources of growth in food production, for example, area expansion versus increased productivity growth (Frink et al. 1999, Tilman et al. 2001, and Cassman et al. 2003). They highlight the environmental loss associated with cropland expansion and focus attention on the continued need for improved productivity (including N fertilizer use) as a means of reducing those pressures. Cassman et al. also argue that there is little scope for agricultural expansion in many parts of the world; so there is, in reality, an even greater urgency to raise crop productivity (particularly of cereals because they occupy such a large share of cropland).

The recent FAO short-term projections to 2008 (FAO 2003) estimated that 92 Mt of fertilizer N would be in use by 2008, but uncertainty about the levels and trends of fertilizer consumption in China may lead to a downward revision of that total (Heffer, personal communication). Clearly, the volatility in fertilizer markets for the past 15 years, coupled with a growth slowdown or reversal in some regions, makes it difficult

Table 18.3. Regional and global projections of fertilizer nitrogen to 2007/2008 2020, and 2050

	2000	2007/2008	N Fertilizer Projections					
			2020			2050		
			IFDC	IMPACT		IFDC	IMPACT	
Regions	FAO-STAT Data	FAO[1] Outlook	Trend + Constant NUE	Food + Constant NUE	Food + Improved NUE	Trend + Constant NUE + Rehab	Food + Constant NUE	Food + Improved NUE
(1)	(2)	(3)	(4)	(5)	(6)	(7)	(8)	(9)
					Million tons			
Africa	2.46	2.85	4.1	3.79	3.68	9.4	4.80	4.40
America	17.20	18.55	21.2	23.26	20.59	30.6	25.1	22.7
North America	12.03	12.55	14.1	15.52	14.30	18.5	16.12	15.17
LAC	5.17	6.01	7.1	7.74	6.28	12.1	8.95	7.54
Asia	45.91	54.75	66.05	62.52	51.68	94.43	69.45	58.93
W. Asia	3.09	3.40	4.6	4.15	3.75	7.6	4.69	4.30
S. Asia	14.55	16.91	23.4	20.19	15.96	35.5	22.78	18.62
E. Asia	28.27	34.44	38.1	38.18	31.97	51.3	41.98	36.01

Eurasia/FSU	2.55	3.03	4.3	3.00	3.01	13.4	3.15	3.15
Europe	11.63	11.57	14.38	16.74	15.33	20.83	16.17	14.98
E. Europe	2.30	2.50	4.1	3.55	2.97	9.9	3.78	3.26
W. Europe	9.33	9.07	10.3	13.19	12.36	10.9	12.39	11.72
Oceania	1.19	1.61	1.6	2.34	1.93	2.6	2.57	2.37
ROW				0.02	0.03	0.03	0.03	
World	80.95	92.35	111.63	111.68	96.24	171.30	121.23	106.56
Developing	52.65			79.46	66.67		89.02	76.34
Developed	28.3			32.22	29.57		32.22	30.23

Source: [1] FAO (2003); other projections made by the authors.

NUE, nitrogen use efficiency; Rehab, rehabilitation of degraded soils; Food based on food projections; FSU, Former Soviet Union; LAC, Latin American countries; ROW, rest of world.

to choose an appropriate base period for projection purposes. Consumption has fluctuated quite significantly in Asia, Eastern Europe, and the former Soviet Union (FSU) in particular.

Most studies defined at least one scenario that addressed the issue of improved NUE, but there was little conceptual or empirical basis to those. The potential increases in NUE assumed in the studies ranged from 15 to 32 percent. The Daberkow et al. study did develop a more structured but still arbitrary means for assessing potential NUE improvements, and this was used with modifications by the authors for the IMPACT-based analysis (reported in columns 6 and 9 of Table 18.3).

Our results in Table 18.3 suggest annual rates of growth in global fertilizer N use at around 1.8 percent in the short term, around 1.6 percent in 2020, and around 1.4 percent in 2050, assuming constant NUE. At the regional level, the fastest growth (sometimes from a low initial level) of 2.0 to 3.5 percent per year is expected in Africa, Oceania, Eastern Europe, and some parts of Central and West Asia. Lowest (and in the short term, sometimes negative) growth rates are anticipated in Western Europe and North America but are expected to stabilize in the longer term in the range 0.5 to 1.2 percent per year (depending on NUE expectations).

Conclusions

Although hard to compare methodologically, successive attempts at projecting N fertilizer use for the medium to long term suggest that future disruption of the global nitrogen balance, even at current levels of NUE, may be less than was once feared, although still of cause for concern and action. It is tricky to assess the reliability of N fertilizer projections because of both method and data shortcomings. Many of the approaches used, and certainly the ones adopted by the authors, are strongly conditioned by recent and projected trends in model determinants. The N fertilizer trend analysis was certainly influenced by the events since the late 1980s in terms of breakup of the FSU and sweeping economic liberalization in China and, to a lesser extent, in many other countries that (among other factors) caused volatility in fertilizer use. For both the trend analysis and the food projection–based analysis, the generally downward adjustments in successive projections of future populations and rates of economic growth also trace through into lower projections for N use compared with earlier assessments. In the food-projection model, however, important and perhaps overoptimistic assumptions were made about our ability to maintain growth in crop productivity. Many concerns are legitimate: underinvestment in publicly funded agricultural research; diminishing exploitable yield gaps in major cereals; overconfidence in the likelihood of biotechnology-based productivity breakthroughs in the short to medium term; soil degradation, salinization, water-logging of irrigated areas, and so on.

Perhaps the most uncertain, yet most critical, assumption on which many of these projections rest is the potential for improving NUE. With loss rates of applied fertilizer

N at around 50 percent, there is, in principle, much scope for improvement; but it is, in truth, highly speculative to make regional and global level predictions about what rates of efficiency might be achieved, over what time scale, and by which farmers. Although effective, low-cost practices to improve nitrogen handling and application efficiency exist, low levels of adoption imply that they do not, as yet, bring tangible benefits to producers.

This brings us to broader questions of data. There is still a great paucity of data on crop-specific nutrient application rates, area fertilized, and corresponding yield responses, notwithstanding the laudable efforts of IFA, IFDC, FAO, and others in compiling the periodic *Fertilizer Use by Crop* publication. Data coverage by both country and crop continues to improve, but there are often insufficient data points to generate region-specific fertilizer response coefficients with acceptable standard errors and not yet any time series of use by crop. It is also frustrating that this data set does not incorporate yield estimates for the specific country, crop, and year to which the fertilizer data correspond.

None of the methods reviewed or used in this study made long-term projections that account for projected market prices of fertilizer. Use of different food models or adaptation of existing models so as to include explicitly fertilizer response into crop production functions might also be a significant step forward, although model parameterization issues at regional and global levels would be significant.

Literature Cited

Bruinsma, J. 2003. *World agriculture: Towards 2015/2030. An FAO perspective.* London: Earthscan Publications Ltd.

Bumb, B., and C. Baanante. 1996. *The role of fertilizer in sustaining food security and protecting the environment to 2020.* Discussion Paper No. 17. Washington, D.C.: International Food Policy Research Institute.

Cassman, K. G., A. Dobermann, D. T. Walters, and H. S. Yang. 2003. Meeting cereal demand while protecting natural resources and improving environmental quality. *Annual Review of Environment and Resources* 28:315–358.

Daberkow, S., K. F. Isherwood, J. Poulisse, and H. Vroomen. 1999. *Fertilizer requirements in 2015 and 2030.* IFA Agricultural Conference on Managing Plant Nutrition. Barcelona: IFA.

FAO (Food and Agriculture Organization of the United Nations). 2003. *Current world fertilizer trends and outlook to 2007/08.* Rome, Italy: FAO.

FAOSTAT. (Food and Agriculture Organization of the United Nations) 2003. *FAO agricultural data bases are obtainable on the world wide web:* http://www.fao.org.

Frink, C. R., P. Waggoner, and J. H. Ausubel. 1999. Nitrogen fertilizer: Retrospect and prospect. *Proceedings of the National Academy of Sciences* 96:1175–1180. http://phe.rockefeller.edu/pnas_nitrogen/pnas_nitrogen.pdf

Galloway, J. N., F. J. Dentener, D. G. Capone, E. W. Boyer, R. W. Howarth, S. P. Seitzinger, G. P. Asner, C. Cleveland, P. Green, E. Holland, D. M. Karl, A. F. Michaels,

J. H. Porter, A. Townsend and C. Vörösmarty. 2004. Nitrogen cycles: Past, present and future. *Biogeochemistry* (in press).

IFA (International Fertilizer Industry Association), IFDC (International Fertilizer Development Center), and FAO (Food and Agriculture Organization of the United Nations). 1999. *Fertilizer use by crop,* 4th ed. Rome, Italy: FAO.

IMF (International Monetary Fund). 2003. *International financial statistics yearbook 2003.* Washington, D.C.: International Monetary Fund.

Rosegrant, M., M. Agcaoili- Sombilla, and N. D. Perez. 1995. *Global food projections to 2020: Implications for investment.* Food Agriculture, and the Environment Discussion Paper No. 5. Washington, D.C.: International Food Policy Research Institute.

Rosegrant, M. W., M. S. Paisner, S. Meijer, and J. Witcover. 2001. *Global food projection to 2020: Emerging trends and alternative futures.* Washington, D.C.: International Food Policy Research Institute.

Rosegrant, M. W., X. Cai, and S. A. Cline. 2002a. *World water and food to 2025: Dealing with scarcity.* Washington, D.C.: International Food Policy Research Institute.

Rosegrant, M. W., S. Meijer, and S. A. Cline. 2002b. *International model for policy analysis of agricultural commodities and trade (IMPACT): Model description.* Washington, D.C.: International Food Policy Research Institute. http://www.ifpri.org/themes/impact /impactmodel.pdf.

Smil, V. 2002. Nitrogen and food production: Proteins for human diets. *Ambio* 31:126–131.

Tilman, D., J. Fargione, B. Wolf, C. D'Antonio, A. Dobson, R. Howarth, D. Schindler, W. H. Schlesinger, D. Simberloff, and D. Swackhamer. 2001. Forecasting agriculturally driven global environmental change. *Scienc*e 292:281–284.

UN (United Nations). 2002. *United Nations millennium declaration.* Resolution adopted by the General Assembly. Document A/RES/55/2. September 2002. New York: United Nations. http://www.un.org/millennium/declaration/ares552e.pdf.

World Bank. 2003. *World development indicators 2003 CD-ROM.* July 2003. ISBN: 0-8213-5423-X. Washington, D.C.: World Bank.

19

Environmental Dimensions of Fertilizer Nitrogen: What Can Be Done to Increase Nitrogen Use Efficiency and Ensure Global Food Security?

Achim Dobermann and Kenneth G. Cassman

Two Sides of the Nitrogen Coin

Human activities have enriched the biosphere with reactive nitrogen (N), resulting in both positive and detrimental effects on ecosystems and human health. Reactive N has been defined as all biologically, photochemically, and radiatively active forms of N—a diverse pool that includes mineral N forms such as NO_3^- and NH_4^+, gases that are chemically active in the troposphere (NO_x and NH_3) and gases such as N_2O that contribute to the greenhouse effect (Galloway et al. 1995). In 1990, the total amount of reactive N created by human activities was about 141 Tg N yr^{-1}, which represents a ninefold increase over the reactive N load in 1890 (Galloway and Cowling 2002). Whereas Asia accounts for nearly 50 percent of the net global creation of reactive N, per capita reactive N load is greatest in North America, followed by Oceania and Europe (Boyer et al., Chapter 16, this volume).

Before 1900, creation of reactive N was dominated by biological N_2 fixation (BNF) in natural ecosystems (Mosier et al. 2001). At present, BNF from cultivated crops, synthetic N production, and fossil fuel combustion are the major sources of reactive N and exceed the contributions from naturally occurring processes (Galloway and Cowling 2002). Fertilizer N contributes about 82 Tg N yr^{-1} reactive N, whereas managed biological fixation adds 20 Tg N yr^{-1} and recycling of organic wastes between 28 to 36 Tg N yr^{-1}. Only about half of all anthropogenic N inputs on croplands are taken up by harvested crops and their residues (Smil 1999). Losses to the atmosphere amount to 26 to 60 Tg N yr^{-1}; ground and surface water bodies receive between 32 to 45 Tg N yr^{-1} from leaching and erosion.

Although such estimates are associated with many uncertainties, it is generally accepted that the current reactive N load is responsible for significant costs to society that occur through direct and indirect negative effects on environmental quality, ecosystem services, and biodiversity. Preliminary estimates for the UK (Pretty et al. 2000) and Germany (Schweigert and van der Ploeg 2000) suggest that the overall environmental costs of N fertilizer use may equal one third of the total value of all farm goods produced. It is not clear, however, that such estimates place an appropriate value on the large *positive* impact of N fertilizer on ensuring food security and adequate human nutrition (Smil 2001) and on the environmental benefits that accrue from avoiding expansion of agriculture into natural ecosystems and marginal areas that cannot sustain crop production.

Producing an adequate supply of human food while protecting environmental quality and conserving natural resources for future generations is the key challenge that must be confronted with regard to N fertilizer use. Improving fertilizer management and the overall N efficiency of cropping systems is a critical component of this challenge because global food security cannot be achieved without meeting the increasing N requirements of crop production (Cassman et al. 2003). Both agronomic and policy actions should target the scales at which major biophysical and socioeconomic variation in the controls on N cycling occurs.

Our principal message is a positive one: Nitrogen use efficiency (NUE) can be increased substantially in most agricultural systems through a combination of (1) better education, (2) adoption of modern management techniques by farmers, (3) continued investment in research and extension, and (4) carefully crafted local policies that contribute to improved N management.

Trends in Nitrogen Fertilizer Use and Nitrogen Use Efficiency

Cereals account for about 64 percent of global N fertilizer use (IFA 2002). Aggregate historical data on global trends in cereal production and fertilizer N consumption have been used to track agriculture's impact on the global N cycle (Tilman et al. 2001). At a global scale, cereal yields and fertilizer N consumption have increased in a near-linear fashion during the past 40 years and are highly correlated with one another (Figure 19.1a). The ratio of global cereal production to global fertilizer N consumption in all crops, a crude index of global NUE in agriculture, has shown a curvilinear decline in the past 40 years, suggesting that future increases in N fertilizer use are unlikely to be as effective in raising yields as in the past (Tilman et al. 2002). Across different countries, the relationship between cereal yields and N use is more scattered (Figure 19.1b)

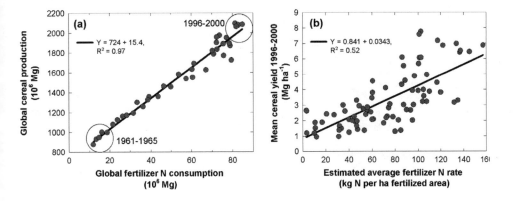

Figure 19.1. (a) Relationship between global cereal production (FAO 2003) and global fertilizer N use on all crops (IFA 2003). Each data point represents 1 year from 1961 to 2000. (b) Relationship between national-level cereal yields and estimated average rates applied to cereal crops. Each data point represents one of 81 countries for which data on fertilizer use by crops were collected through surveys and expert opinions (IFA 2002).

although still significant. To some degree, the greater scatter is caused by inaccurate estimates of N use by crop, but Figure 19.1b illustrates mainly that the tight linear global relationship between cereal production and N consumption (Figure 19.1a) cannot be generalized to regional, national, or field scales. Historical trends differ widely among regions and countries, and crop yield response to N varies widely among different environments, with most of the variation occurring at the field scale (Cassman et al. 2003).

Nitrogen use in Asia has increased nearly 17-fold since 1961. It rose steeply during the course of the Green Revolution (Figure 19.2a), mainly because of the rapid adoption of modern high-yielding rice and wheat varieties. Large relative increases in N use have also occurred in Latin America (11-fold) and Africa (sevenfold) but starting from very low levels. In Western Europe and North America, N fertilizer use has remained relatively constant during the past 25 years or has slightly declined, whereas yields of many crops continue to rise. In Eastern Europe and the countries of the Former Soviet Union (FSU), N consumption dropped sharply in the 1990s as a result of political and economic turmoil.

These same regions also show different trends in the ratio of cereal production to N fertilizer consumption of all crops, which can be considered a crude indicator of NUE at the national or regional scale (Figure 19.2b). In general, large values for this ratio are typical of low-input systems that use little N fertilizer, whereas intensified cropping systems with high input levels tend to have small values. The ratio was already low in North

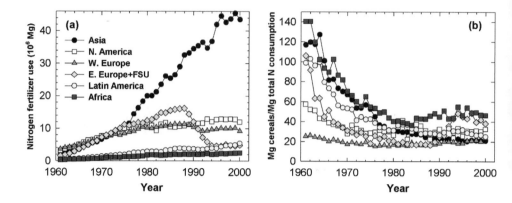

Figure 19.2. Trends in regional consumption of N fertilizer (IFA 2003) applied to all crops (a) and the ratio of cereal production (FAO 2003) to total N fertilizer consumption in selected world regions (not shown: Near East and Oceania). (b) Total N consumption in these six regions represents 95 percent of the global fertilizer N use. Eastern Europe and the former Soviet Union (FSU) includes the 14 countries of central and Eastern Europe and all 15 countries of the FSU. Note that historical changes in the share of N use by cereal crops may have affected some of the trends.

America and Western Europe during the early 1960s and bottomed out in 1980 at about 15 to 30 Mg grain Mg N^{-1}. Since then, it has gradually increased as a result of crop improvement and adoption of better management technologies. In Asia, a sharp decline occurred from more than 100 Mg grain Mg N^{-1} in the early 1960s to about 20 Mg grain Mg N^{-1} since 1995. In Eastern Europe and the FSU, declining N use led to increases in the ratio since 1990, and values are now similar to those observed in Africa (40-50 Mg grain Mg N^{-1}). In general, where the regional use of fertilizer N has increased, the ratio of cereal production to N consumption has approached similar low levels; where N use is low, the ratio tends to be twice as large as in regions with high N use, but crop productivity is low.

Further differentiation occurs when trends of individual countries are compared. The eight countries shown in Figure 19.3 account for 63 percent of the global fertilizer N consumption. In four industrialized countries with intensive agriculture supported by sophisticated infrastructure (the United States, Germany, UK, Japan), N use has either stagnated or substantially declined since 1980, even though except for Japan cereal yields have continued to increase at about the same pace as before 1980 (Figure 19.3a and c). Overall, the ratio of cereal production to national N fertilizer use has begun to increase in these industrialized countries (Figure 19.3e). In four developing countries where agricultural systems continue to intensify, N fertilizer use has increased dramatically since the mid 1960s (Figure 19.3b), which has contributed much to the increases in crop yields (Figure 19.3d). In all four countries, however, average cereal yields remain well

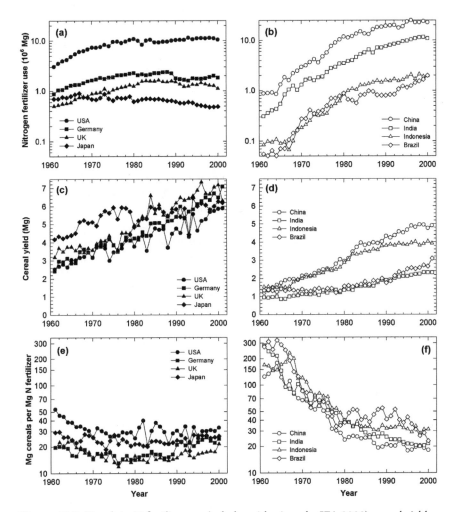

Figure 19.3. Trends in N fertilizer use (a, b; logarithmic scale; IFA 2003), cereal yields (c, d: FAO 2003), and the ratio of national cereal production to national N use (e, f; logarithmic scale) in selected countries. Left: Four developed countries of the Western world. Right: Four transition countries with rapid adoption of N fertilizers as part of crop intensification taking place during the past 30 to 40 years.

below yield levels in the four industrialized countries, and the ratio of cereal production to N use continues to decline at about the same pace (Figure 19.3f).

Maize production in the United States provides a specific example of factors that affect national trends in N use. In U.S. maize systems, NUE increased from 42 kg grain kg N^{-1} in 1980 to 57 kg grain kg N^{-1} in 2000. Factors that contributed to this improvement included (1) increased yields and more vigorous crop growth associated with greater stress tolerance of modern maize hybrids (Duvick and Cassman 1999); (2) improved management of production factors other than N (conservation tillage, seed quality, higher plant densities, weed and pest control, balanced fertilization with other nutrients, irrigation); and (3) improved N fertilizer management to better match the amount and timing of applied N to crop N demand and the N supply from indigenous resources (Dobermann and Cassman 2002).

Large investments in research were made, and adoption of improved technologies by farmers required additional investments in extension education. Legislative measures to control N use were implemented only in selected areas where groundwater contamination or runoff pollution exceeded thresholds that negatively affected water quality. Japan is another example for how social factors as well as investments in research and extension can change national trends in N use and NUE. In Japan, cereal yields have increased little for the past 25 years (Figure 19.3c), but national fertilizer N use has declined substantially (Figure 19.3a). The shift to rice varieties with higher grain quality and adoption of more knowledge-intensive N management practices were the key factors responsible for this decrease in N use (Suzuki 1997). As a result, NUE of rice has increased from 57 kg grain kg N^{-1} in 1985 to more than 75 kg grain kg N^{-1} in recent years (Mishima 2001).

Lessons learned from these regional and national trends are (1) it is dangerous to extrapolate past global trends into the future (e.g., Tilman et al. 2001) because aggregate global data do not provide a sound basis for assessing regional and local mitigation options or the potential to improve NUE; (2) global N policies must account for significant regional and national differences in the intensity of N use as well as in the sophistication of crop management technologies utilized to manage N and other factors that influence NUE; (3) an initial decline in NUE is often inevitable and not necessarily bad in cropping systems that undergo rapid intensification that leads to substantial increases in food production; and (4) even at high production levels, NUE can be increased by adoption of improved management practices provided that substantial investments are made in research and extension.

Because the relationship between crop yield and N uptake is tightly conserved, achieving further increases in world food production will require greater N uptake by crops and, consequently, either more N fertilizer or more efficient use of applied N. Estimates of future growth in global N consumption differ because of different forecasting methods and assumptions about food demand, land area, yields, nutrient sources and trends in NUE (Wood et al., Chapter 18, this volume). For example, recent estimated increases in the global N fertilizer requirements for rice, wheat, and maize to 2025 ranged from 19 to 61 percent to produce adequate supplies of these cereals (Cassman et al. 2003). In the most probable scenario, current trends of a slight decrease in the harvested area of cereals may continue, but NUE could be increased by 15 percent relative to present levels. If so, achieving the required 37 percent increase in cereal production would require only a 19 percent increase in N fertilizer use (Cassman et al. 2003). Not included in this were potential decreases in N fertilizer requirements resulting from replacement with organic N sources, such as manure or biological N fixation. Globally, it is unlikely that there is enough land or organic N materials to support such a replacement. The contribution made by manure relative to fertilizers plus manure has declined from 60 percent in 1961 to about 25 percent at present, although important regional differences in the importance of manure exist (Sheldrick et al. 2003). In some developed countries with large livestock industries, fertilizer demand may decrease because of an increase in manure availability. In some developing countries with little access to fertilizers, manure and other organic nutrient sources may remain the major nutrient inputs.

What Can Be Done to Increase Nitrogen Use Efficiency?

Conceptual Framework Used by Agronomists

Both past trends and future projections for N use illustrate the need to increase NUE as a key control on the amount of reactive N cycling through global ecosystems. Agricultural lands lose a substantial fraction of the fertilizer N applied, often 40 to 70 percent. Although examples of increasing NUE have been documented, even at national scales (Figure 19.3), current average levels of NUE remain well below those that could be achieved with improved technologies (Cassman et al. 2002). Raising NUE requires action at the field scale and a thorough understanding of the environment and management effects on the N cycle at this scale.

In an individual field or experimental plot, grain yield (Y) and plant N accumulation (U) increase with increasing N rate (F) and gradually approach a ceiling (Figures 19.4a and c). The level of this ceiling is determined by the site yield potential. At low levels of N supply, rates of increase in yield and N uptake are large because N is the primary factor limiting crop growth and final yield. As the N supply increases, incremental yield gains become smaller because yield determinants other than N become more limiting as the maximum yield potential is approached. The broadest measure of NUE is the ratio of yield to the amount of applied N, also called the *partial factor productivity* (PFP_N) of applied N, which declines with increasing N application rates (Figure 19.4a). The PFP_N is an aggregate efficiency index that includes contributions to crop yield derived from uptake of indigenous soil N, N fertilizer uptake efficiency, and the efficiency with which N acquired by the plant is converted to grain yield.

The PFP_N has limited potential for identifying specific constraints to improving N efficiency and the most promising management strategies to alleviate these constraints. Many agronomists therefore use a framework (see Appendix) in which NUE of a single crop is separated into different component indices (Cassman et al. 1996; Novoa and Loomis 1981). The incremental yield increase that results from N application is defined

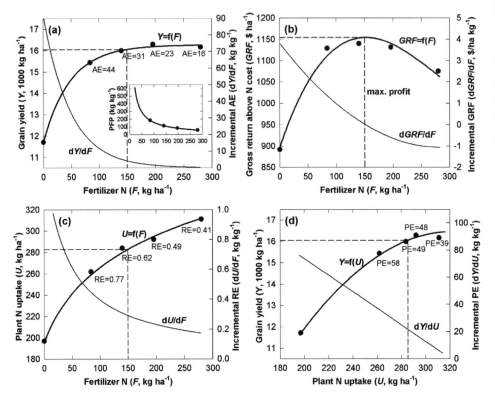

Figure 19.4. Response of irrigated maize to N application at Clay Center, Nebraska, 2002. (a) Relationship between the grain yield (Y) and the N rate (F) and the incremental agronomic N efficiency (AE, kg of grain yield increase per kg of N applied). (b) Relationship between gross return above fertilizer cost (GRF) and the N rate and the incremental GRF (dGRF/dF). (c) Relationship between plant N accumulation (U) and N rate and the incremental recovery efficiency of fertilizer N (RE, kg increase in N uptake per kg N applied). (d) Relationship between grain yield and plant N accumulation (U) and the incremental physiologic efficiency of fertilizer N (PE, kg increase in grain yield per kg N taken up). In all graphs, the dashed lines indicate where maximum profit occurred. Measured values of AE (a), RE (c), and PE (d) calculated by the difference method are shown for the four rates of N application. The insert in graph (a) shows the decline in the partial factor productivity of fertilizer N (PFP, ratio Y/F) with increasing N rate (Source: Nebraska Soil Fertility Project; data collected by R. Ferguson, University of Nebraska—Lincoln).

as the *agronomic efficiency* (AE_N) of applied N, which itself is the product of the recovery efficiency of N (RE_N: the percentage of fertilizer N recovered in aboveground plant biomass during the growing season) and the efficiency with which the plant uses each unit of N acquired from applied N to produce grain (PE_N). These techniques provide "snapshots" of efficiency estimates for given levels of N application (Figures 19.4a, c, and d). In addition, continuous response functions between yield, plant N uptake, and fertilizer N input can be used to illustrate the curvilinear nature of crop response to N application. They also allow establishing economically optimum levels of N use (Figure 19.4b) and the first derivatives of the fitted curves represent the incremental changes in AE_N (Figure 19.4a), RE_N (Figure 19.4c), and PE_N (Figure 19.4d) at a specific N fertilizer rate.

The example shown in Figure 19.4 is notable because it illustrates a crop N response for soil with a large indigenous N supply in an environment with high-yield potential, including good overall crop and soil management. A modern maize hybrid was grown with full irrigation, balanced supply of nutrients, and excellent weed and pest control. Grain yield without applied N was 11.7 Mg ha^{-1}, and yield increased to more than 16 Mg ha^{-1} with N application, a yield level close to the simulated genetic–climatic yield potential. Maximum profit was reached at an N rate of 150 kg ha^{-1}, which was close to one of the actual N treatments used in the experiment (144 kg ha^{-1}). At that level of applied N, AE_N and RE_N estimated by the difference method were high, 31 kg grain kg N^{-1} and 62 percent, respectively; but the curvilinear relationship between AE_N or RE_N as a function of N (Figures 19.4a and c) makes the incremental increase in AE_N and RE_N at 144 kg applied N ha^{-1} only 5 kg grain kg N^{-1} and 30 percent, respectively. Nitrogen fertilizer applied in excess of this rate would be used very inefficiently.

Apparent Disconnection of Fertilizer Nitrogen Use and Crop Yields at the Farm Level

Krupnik et al. (Chapter 14, this volume) provide a summary of the literature on N recovery by crops when evaluated at different spatial scales. What is obvious from their analysis is that there is a paucity of reliable data on NUE based on measurements from on-farm studies in the major crop production systems. Likewise, we are not aware of measurements of on-farm NUE that include the contributions from both RE and changes in soil N reserves. This shortage of information reflects the logistical difficulty and high cost of obtaining direct on-farm measurements and the lack of funding for what appear to be routine on-farm evaluations (Cassman et al. 2002). Instead, agronomic research appears to focus on short-term studies at few research sites, with insufficient geographical context, little use of spatial information, and only scarce application of modeling tools to allow extrapolation of the results (White et al. 2002).

The picture emerging from recent on-farm studies is one of an apparent disconnection between the amount of fertilizer N applied by farmers and the crop yield that is achieved, resulting in often low and highly variable NUE. Rice systems in Asia and Africa have been investigated most with regard to on-farm measurement of NUE following a robust methodology applied in numerous on-farm studies for nearly 10 years (Cassman et al. 1996; Dobermann et al. 2004b). The following were the major conclusions:

1. Large spatial and temporal variability exists among fields with regard to indigenous N supply, fertilizer use, crop yields, NUE, and marginal return from N fertilizer.
2. Grain yield obtained by farmers is closely correlated with plant N uptake, but not with fertilizer N use; NUE is often not related to N rates or the supply of N from indigenous sources.
3. Climate, the supply of other essential nutrients, disease, insect pest, and weed pressure, stand establishment, water management, and N management technology (timing, forms, placement, etc.) have large effects on RE_N and PE_N and, therefore, the overall crop response to N fertilizer.
4. It is difficult to predict the N supply from indigenous sources using existing assessment methods such as soil tests because they fail to account for the dynamics of nutrient supply, including N provided from a range of indigenous sources including soil organic matter, irrigation, or biological N_2 fixation.

Extensive on-farm studies of similar kind and nearly global scope have not been conducted in other environments or for other major cereal crops, which makes it difficult to judge whether the findings made for rice systems are applicable to other crops and cropping systems. Some evidence exists, however, that this may be the case for wheat grown in rice–wheat systems of south Asia and maize grown in rain-fed and irrigated systems of the U.S. Corn Belt (Adhikari et al. 1999; Cassman et al. 2002).

New Farm-level Nitrogen Management Strategies

Cassman et al. (2002) concluded that the average RE_N ranges from about 20 to 50 percent in major cereal cropping systems of the world, whereas levels of 60 to 80 percent are commonly achieved with excellent management in research trials (Krupnik et al., Chapter 14, this volume). We do not believe the gap between actual average RE_N achieved by farmers and the potential RE_N with improved management is due to differences in N accumulation in soil organic matter because most of the major cereal production systems are most likely to be at a relatively steady-state with regard to soil C and N sequestration. Hence, this discrepancy is usually explained by "scale effects": Small research plots can be managed more accurately with regard to operations such as tillage, seeding, nutrient applications, weed and pest management, irrigation, and harvest, which all affect efficiency. At issue, therefore, is how farm-scale technologies can be improved to enable farmers to achieve NUE similar to the highest efficiency levels measured in small research plots. If the gap between current on-farm NUE and the potential NUE achieved in research plots could be significantly reduced, the impact on minimizing the potential negative effects of N fertilizer use on environmental quality could be greatly reduced. Indeed, increased cereal yields could be achieved without a large increase in the total amount of N fertilizer use (Cassman et al. 2003).

Because the relationship between crop yield and N supply follows a diminishing return function (Figure 19.4), achieving high yields and high NUE without increasing N losses is a difficult task. Most farmers could easily increase NUE by applying less N. Where fertilizer use is in excess of crop needs, this is a viable strategy. Otherwise, loss of yield and income will occur, which is why farmers often apply a certain amount of "insurance N" in excess of the optimum rate. Research literature on improving NUE in crop production systems has emphasized the need for greater synchrony between crop N demand and the N supply from all sources throughout the growing season (Appel 1994; Cassman et al. 2002). This approach explicitly recognizes the need to use efficiently both indigenous and applied N, which is justified by the fact that N losses increase in proportion to the amount of available N present in the soil profile at any given time. Profit and RE_N are optimized with the least possible N losses when the plant-available N pool is maintained at the minimum size required to meet crop N requirements at each stage of growth.

Increased ability to acquire and utilize N, better fertilizers, and better N management strategies of different cropping systems will improve NUE (Giller et al., Chapter 3, this volume). Major efforts have concentrated on integrated nutrient management, providing better N prescriptions (e.g., better N splitting schemes or by managing spatial variability through precision farming) or by managing the dynamics of soil N supply and crop N demand (e.g., through real-time N management, modified fertilizers, inhibitors, or placement techniques that avoid excessive accumulation of mineral N in the soil). Prerequisites for implementing such approaches in practice are that they must be simple, involve little extra time, provide consistent gains in NUE and yield, and are cost-effective.

Many of the new products or techniques are not yet widely used, often for cost reasons. In large-scale agriculture practiced in developed countries, for example, precision farming studies have demonstrated that variable-rate N fertilizer application can significantly reduce the N rate required (often by about 10–30 percent) to achieve yields similar to those obtained with standard uniform management (Dobermann et al. 2004a). The management tools used in these studies varied widely. Initial work focused on soil and yield mapping as the basis for prescribing spatially variable N rates. More recently, emphasis has shifted toward real-time methods of N management that utilize crop simulation models, remote sensing, or on-the-go crop sensing/variable-rate N spreaders to determine the spatially variable needs for N at critical growth stages. On-the-go sensing of crop "greenness" to control N applications in cereals has recently been commercialized in parts of Europe and the United States. It remains to be seen, however, whether these technologies can be made cost-effective because gains in yield and profitability tend to be small. It is also not yet clear whether precision farming approaches can result in measurable decreases in nitrate leaching risk (Ferguson et al. 2002) or N_2O emissions.

Increasing NUE in the developing world presents similar challenges but also great opportunities. Site-specific management in irrigated rice systems of Asia has focused on managing nutrients at the scale of a single small field, including in-season N management decisions. One line of research has focused on corrective, in-season N management using tools such as a chlorophyll meter (Peng et al. 1996). During the growing season, N fertilizer is applied whenever the leaf N status falls below an empirically calibrated threshold. This same approach can be followed using a simple leaf color chart (LCC) if local calibrations have been established (Yang et al. 2003). Evaluation of these N management methods has generally shown that the same rice yield can be achieved with about 20 to 30 percent less N fertilizer applied, whereas increases in yield appear to be less common or are relatively small (Giller et al., Chapter 3, this volume). There are, however, risks involved: leaf color can be affected by growth limitation other than N deficiency; the decision on when and how much N to apply remains empirical and difficult; periods of N deficiency may occur in between diagnosis events; decisions about early season applications of N and other nutrients must be made using other methods; sampling and measurement errors can occur; quality control in making the LCC must assure reproducible color hues.

Some of these uncertainties were addressed in a broader site-specific nutrient management (SSNM) concept (Dobermann et al. 2004b). Key components of this approach were measurement of grain yield in nutrient omission plots to obtain field-specific estimates of the indigenous supply of N, P, and K; a decision support system for predicting nutrient requirements and the optimal amount of N to be applied before planting; and in-season upward or downward adjustments of predetermined N topdressings at critical growth stages based on chlorophyll meter or LCC readings. From 1997 to 2000, this SSNM strategy was evaluated in 179 farmer's fields in eight major irrigated rice areas of Asia (Dobermann et al. 2002). On average, grain yield increased by about 11 percent and the N fertilizer rate decreased by 4 percent compared with the baseline farmers' fertilizer practice. Average profit increased by US\$46 ha^{-1} per crop cycle. The SSNM approach increased the probability of obtaining a greater RE_N, which indicated consistent improvement in efficiency across farms and production domains (Figure 19.5). Average RE_N increased from 31 percent to 40 percent, with 20 percent of all farmers achieving more than 50 percent RE_N. The two approaches described above were recently integrated into a flexible framework of simple SSNM principles for rice (Witt et al. 2004).

Examples of achieving increased NUE in a cost-effective manner have also been documented in other cropping systems and environments—often involving much simpler approaches than the SSNM technologies described above. Improving the congruence between crop N demand and N supply through better fine-tuning of split applications increased N fertilizer efficiency of irrigated wheat in Mexico (Riley et al. 2003). In Mauritania, improved nutrient and weed management recommendations increased NUE and resulted in large additive effects on yields and profitability of irrigated rice (Haefele et al. 2001). In Nepal, simply following the existing nutrient recommendations increased wheat yields at 21 locations by 40 percent at an average RE_N of 52 percent (Adhikari et al. 1999).

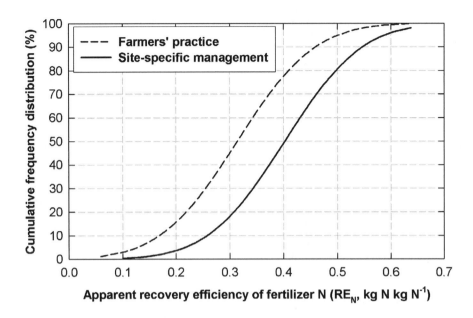

Figure 19.5. Shift in the cumulative frequency distribution of the apparent recovery efficiency of N (RE_N; kg increase in N uptake per kg N applied) resulting from site-specific nutrient management in irrigated rice. Cumulative distribution functions are based on RE_N values measured for four consecutive rice crops grown in 179 farmers' fields in China, India, Indonesia, Thailand, the Philippines, and Vietnam from 1997 through 1999. Monitoring in each field and growing season included replicated N omission plots, sampling plots in the farmers' fertilizer practice, and permanent plots that were managed with a site-specific nutrient management approach (Dobermann et al. 2002).

In summary, there is significant potential to increase NUE at the farm level, and many of the concepts and tools needed to achieve such an increase have already been developed. A key challenge is to ensure that these technologies are cost-effective and user-friendly such that they are attractive options for adoption by farmers. Besides the socioeconomic factors involved, scientists must also solve the puzzle of how to produce more yield with less N per unit of grain yield. Many of the currently used decision tools involve empirical calibration (Schroeder et al. 2000), whereas more significant gains should be possible from more quantitative approaches of characterizing N needs in relation to site yield potential and other crop management factors (Dobermann and Cassman 2002).

Strategic, interdisciplinary field research is required to understand the upper limits to yield and resource use efficiency in a particular environment, with specific attention to identifying how management factors interact to influence crop performance and NUE. Table 19.1 illustrates this for a maize experiment at Lincoln, Nebraska. High

Table 19.1. Trends in nitrogen use, maize grain yield, and partial factor productivity of fertilizer-nitrogen (PFP_N) in a high-yielding field experiment at Lincoln, Nebraska (2000–2003). Recommended management: Normal plant density and existing best nutrient management practice; intensive management: High plant density and intensive nutrient management aiming at yields near the yield potential

Treatment		Mean	2000	2001	2002	2003
Continuous maize						
Recommended	Fertilizer N (kg ha^{-1})	191	203	200	180	180
	Grain yield (Mg ha^{-1})	13.6	13.4	14.0	11.2	16.0
	PFP_N (kg grain kg N^{-1})	72	66	70	62	89
Intensive[1]	Fertilizer N (kg ha^{-1})	301	363	300	289	250
	Grain yield (Mg ha^{-1})	15.5	14.4	15.8	15.2	16.6
	PFP_N (kg grain kg N^{-1})	53	40	53	53	66
Maize following soybean						
Recommended	Fertilizer N (kg ha^{-1})	130	138	130	120	130
	Grain yield (Mg ha^{-1})	14.8	14.1	14.4	13.9	16.8
	PFP_N (kg grain kg N^{-1})	115	102	111	116	129
Intensive	Fertilizer N (kg ha^{-1})	251	298	240	216	250
	Grain yield (Mg ha^{-1})	16.1	15.6	15.6	15.3	17.9
	PFP_N (kg grain kg N^{-1})	65	52	65	71	72

[1] 2002 and 2003 include application of some N in previous fall to support residue decomposition.

maize yields and high NUE were consistently achieved over several years either by following existing best management practices or by gradually fine-tuning management in a high-yield system, particularly with regard to the evolving N requirements. In the two treatments representing current best management practices yields were 60 to 70 percent greater than state or national averages (8.5–9 Mg ha^{-1}), and PFP_N was also significantly larger than the current state average (55 kg grain kg N^{-1}). In corn following soybean, grown with recommended best management practices, average PFP_N was 115 kg grain kg N^{-1}, a level at which RE_N averaged about 80 percent. In the two intensively managed systems with much larger N input, but also more splitting of N applications, yields approached the site yield potential, but PFP_N was also gradually increased over time and exceeded the average NUE achieved by Nebraska maize farmers (Table 19.1). Changes in the indigenous soil N supply were caused by increasing soil organic matter content, leading to adjustments in N rates and significant increases in PFP_N over time.

Research and Policy Implications

Agriculture can make a substantial contribution to reducing the global reactive N load, but improving NUE in major food crops requires collaboration among agronomists, soil scientists, agricultural economists, sociologists, ecologists, and politicians (Galloway et al. 2002). Reducing the reactive N load will require increases in PFP_N through increasing both the indigenous N supply and RE_N, which in turn will require innovative crop and soil management practices. The economic benefit-to-cost ratio has a large influence on farmer adoption of new technologies. Whereas some management practices might increase PFP_N by reducing N losses or increasing the proportion of N inputs that are retained in soil organic and inorganic N pools, adoption by farmers is not likely without the promise of adequate economic return or incentives that may help the adoption of new techniques. Hence, management options must also consider RE_N and PE_N because these parameters determine the economic impact on grain yield in relation to applied N inputs and crop N accumulation.

The examples described here demonstrate that there is much potential for fine-tuning N management to increase NUE. Increases in RE_N of about 30 percent relative to present levels appear feasible in many environments. Such improvements are likely to have large impact on the rates, sources, and sinks in the global N cycle, but they require suitable policies and significant long-term investments in research and extension education. Implementing global or regional policies on N use in agriculture is difficult because of different agricultural priorities in different countries (Mosier et al. 2001). Policies that simply promote an increase or decrease in N fertilizer use at a national or state level would have a widely varying impact on yields, farm profitability, NUE, and environmental quality. Given the scattered nature of the relationships between N use, yield, and NUE at the farm level, restrictions on N use across the board would penalize farmers unevenly. Good farmers who are efficient in their use of N fertilizer and operate near the upper threshold of potential NUE will lose yield and income if blanket regulations on N use are enforced. Farmers who are poor field managers may not lose yield and achieve higher net returns from a forced reduction in N fertilizer rate because factors other than N limit yields in their fields. Instead, achieving greater NUE at national, regional, and global scales can be achieved only with policies and investments in research and extension that target increases in NUE at the field scale.

Current investments to support research on technologies that can achieve greater congruence between crop N demand and N supply from all sources—including fertilizer, organic inputs, and indigenous soil N—are insufficient given the need to sustain the rate of yield increases to meet food demand without a major expansion in cultivated area (Cassman et al. 2002). Optimizing the timing, quantity, and availability of applied N is the key to achieving high RE_N and to increasing AE_N and the overall NUE. Many new technologies have not been adopted because savings in fertilizer N may not be cost-efficient or because the technologies themselves were too knowledge or labor intensive. Support is needed for developing robust technologies and the supporting infrastructure, including incentives for the use of advanced N management practices rather than forced reductions in N rates applied to all farmers without regard to the NUE they currently achieve and their potential contribution to the reactive N load in off-farm environments.

Reliable estimates of N losses from the major agroecosystems are required to understand the contribution of agriculture to the environmental problems caused by too much reactive N in the environment. Too few studies have been done in which N losses have been measured in on-farm settings across a reasonable range of representative environments and spatial scales, including watershed-based measurements; most estimates are based on field experiments conducted at research stations. A more comprehensive assessment of NUE in the major crop production systems would help identify which systems should receive greatest attention with regard to improving NUE.

Adoption of improved technologies requires additional skills and labor or investments in new equipment. Information on expected costs and economic returns from such investments is required to convince farmers of the benefits from adoption. The only data directly available to farmers regarding NUE are the grain yield they obtain from their fields and the amount of N fertilizer they apply. Unfortunately, these data provide little information about the size of the indigenous N supply, RE_N, or PE_N, all of which are essential for identifying management practices that increase both NUE of the cropping system and economic return from applied N. Farmers also need estimates of the portion of yield obtained from indigenous soil N and the yield increase from applied N. A more thorough understanding of these NUE components is essential for management decisions that maximize returns from both indigenous and applied N and that in turn minimize the potential for N losses.

Literature Cited

Adhikari, C., K. F. Bronson, G. M. Panaullah, A. P. Regmi, P. K. Saha, A. Dobermann, D. C. Olk, P. Hobbs, and E. Pasuquin. 1999. On-farm soil N supply and N nutrition in the rice-wheat system of Nepal and Bangladesh. *Field Crops Research* 64:273–286.

Appel, T. 1994. Relevance of soil N mineralization, total N demand of crops and efficiency of applied N for fertilizer recommendations for cereals—Theory and application. *Zeitschrift fur Pflanzenernaehrung und Bodenkunde* 157:407–414.

Cassman, K. G., A. Dobermann, and D. T. Walters. 2002. Agroecosystems, nitrogen-use efficiency, and nitrogen management. *Ambio* 31:132–140.

Cassman, K. G., A. Dobermann, D. T. Walters, and H. S. Yang. 2003. Meeting cereal demand while protecting natural resources and improving environmental quality. *Annual Review of Environment and Resources* 28:315–358.

Cassman, K. G., H. C. Gines, M. Dizon, M. I. Samson, and J. M. Alcantara. 1996. Nitrogen-use efficiency in tropical lowland rice systems: Contributions from indigenous and applied nitrogen. *Field Crops Research* 47:1–12.

Dobermann, A., B. S. Blackmore, S. E. Cook, and V. I. Adamchuk. 2004a. Precision farming: Challenges and future directions, in *New directions for a diverse planet. Proceedings of the 4th International Crop Science Congress, 26 Sept–1 Oct 2004, Brisbane, Australia [CD-ROM]*, Brisbane.

Dobermann, A., and K. G. Cassman. 2002. Plant nutrient management for enhanced productivity in intensive grain production systems of the United States and Asia. *Plant and Soil* 247:153–175.

Dobermann, A., C. Witt, and D. Dawe. 2004b. *Increasing productivity of intensive rice systems through site-specific nutrient management*. Enfield, New Hampshire: Science Publishers, Inc., and Los Baños (Philippines): International Rice Research Institute.

Dobermann, A., C. Witt, D. Dawe, G. C. Gines, R. Nagarajan, S. Satawathananont, T. T. Son, P. S. Tan, G. H. Wang, N. V. Chien, V. T. K. Thoa, C. V. Phung, P. Stalin, P. Muthukrishnan, V. Ravi, M. Babu, S. Chatuporn, M. Kongchum, Q. Sun, R. Fu, G. C. Simbahan, and M. A. A. Adviento. 2002. Site-specific nutrient management for intensive rice cropping systems in Asia. *Field Crops Research* 74:37–66.

Duvick, D. N., and K. G. Cassman. 1999. Post-green revolution trends in yield potential of temperate maize in the North-Central United States. *Crop Science* 39:1622–1630.

FAO (Food and Agricultural Organization of the United Nations). 2003. *FAOSTAT Database—Agricultural Production*. http://apps.fao.org. Rome, Italy: FAO.

Ferguson, R. B., G. W. Hergert, J. S. Schepers, C. A. Gotway, J. E. Cahoon, and T. A. Peterson. 2002. Site-specific nitrogen management of irrigated maize: Yield and soil residual nitrate effects. *Soil Science Society of America Journal* 66:544–553.

Galloway, J. N., and E. B. Cowling. 2002. Reactive nitrogen and the world: 200 years of change. *Ambio* 31:64–71.

Galloway, J. N., E. B. Cowling, S. P. Seitzinger, and R. H. Socolow. 2002. Reactive nitrogen: Too much of a good thing? *Ambio* 31:60–63.

Galloway, J. N., W. H. Schlesinger, H. Levy, A. Michaels, and J. L. Schnoor. 1995. Nitrogen fixation: Atmospheric enhancement—environmental response. *Global Biochemical Cycles* 9:235–252.

Haefele, S. M., M. C. S. Wopereis, C. Donovan, and J. Maubuisson. 2001. Improving the productivity and profitability of irrigated rice production in Mauritania. *European Journal of Agronomy* 14:181–196.

IFA (International Fertilizer Industry Association). 2002. *Fertilizer use by crop*. Rome, Italy: IFA, International Fertilizer Development Centre (IFDC), International Potash Institute (IPI), Potash and Phosphate Institute (PPI), Food and Agriculture Organization (FAO).

IFA (International Fertilizer Industry Association). 2003. *IFADATA statistics*. http://www.fertilizer.org/ifa/statistics.asp. Paris, France: IFA.

Mishima, S. 2001. Recent trend of nitrogen flow associated with agricultural production in Japan. *Soil Science and Plant Nutrition* 47:157–166.

Mosier, A. R., M. A. Bleken, P. Chaiwanakupt, E. C. Ellis, J. R. Freney, R. B. Howarth, P. A. Matson, K. Minami, R. Naylor, K. N. Weeks, and Z. L. Zhu. 2001. Policy implications of human-accelerated nitrogen cycling. *Biogeochemistry* 52:281–320.

Novoa, R., and R. S. Loomis. 1981. Nitrogen and plant production. *Plant and Soil* 58:177–204.

Peng, S., F. V. Garcia, R. C. Laza, A. L. Sanico, R. M. Visperas, and K. G. Cassman.

1996. Increased N-use efficiency using a chlorophyll meter on high-yielding irrigated rice. *Field Crops Research* 47:243–252.

Pretty, J., C. Brett, D. Gee, R. E. Hine, C. F. Mason, J. I. L. Morison, H. Raven, M. D. Rayment, and G. van der Bijl. 2000. An assessment of the total external costs of UK agriculture. *Agricultural Systems* 65:113–136.

Riley, W. J., I. Ortiz-Monasterio, and P. A. Matson. 2003. Nitrogen leaching and soil nitrate, nitrite, and ammonium levels under irrigated wheat in Northern Mexico. *Nutrient Cycling in Agroecosystems* 61:223–236.

Schroeder, J. J., J. J. Neeteson, O. Oenema, and P. C. Struik. 2000. Does the crop or the soil indicate how to save nitrogen in maize production? Reviewing the state of the art. *Field Crops Research* 66:151–164.

Schweigert, P., and R. R. van der Ploeg. 2000. Nitrogen use efficiency in German agriculture since 1950: Facts and evaluation. *Berichte über Landwirtschaft* 80:185–212.

Sheldrick, W. F., J. K. Syers, and J. Lingard. 2003. Contribution of livestock excreta to nutrient balances. *Nutrient Cycling in Agroecosystems* 66:119–131.

Smil, V. 1999. Nitrogen in crop production: An account of global flows. *Global Biochemical Cycles* 13:647–662.

Smil, V. 2001. *Enriching the earth: Fritz Haber, Carl Bosch, and the transformation of world food production.* Cambridge, Massachusetts: The MIT Press.

Suzuki, A. 1997. *Fertilization of rice in Japan.* Tokyo, Japan: Japan FAO Association.

Tilman, D., K. G. Cassman, P. A. Matson, R. L. Naylor, and S. Polasky. 2002. Agricultural sustainability and intensive production practices. *Nature* 418:671–677.

Tilman, D., J. Fargione, B. Wolff, C. D'Antonio, A. Dobson, R. W. Howarth, D. Schindler, W. H. Schlesinger, D. Simberloff, and D. Swackhamer. 2001. Forecasting agriculturally driven global environmental change. *Science* 292:281–284.

White, J. D., J. D. Corbett, and A. Dobermann. 2002. Insufficient geographic characterization and analysis in the planning, execution and dissemination of agronomic research?. *Field Crops Research* 76:45–54.

Witt, C., R. J. Buresh, V. Balasubramanian, D. Dawe, and A. Dobermann. 2004. Principles and promotion of site-specific nutrient management. Pp. 397–410 in *Increasing productivity of intensive rice systems through site-specific nutrient management,* edited by A. Dobermann, C. Witt, and D. Dawe. Enfield, New Hampshire: Science Publishers and Los Baños (Philippines): International Rice Research Institute.

Yang, W. H., S. Peng, J. Huang, A. L. Sanico, R. J. Buresh, and C. Witt. 2003. Using leaf color charts to estimate leaf nitrogen status of rice. *Agronomy Journal* 95:212–217.

Appendix

Agronomic Indices of Nitrogen Use Efficiency

The following indices are widely used in research on assessing the efficiency of applied N and are independent of scale. These indices are used mainly for purposes that emphasize crop response to fertilizer N. They are rarely used in systems where organic sources and biological N fixation are the major N inputs.

PFP_N = partial factor productivity from applied N (kg product kg^{-1} N applied)
= Y_N/F_N

where Y_N is the crop yield (kg ha^{-1}) at a certain level of fertilizer N applied (F_N, kg ha^{-1}).

AE_N = agronomic efficiency of applied N [kg product increase kg^{-1} N applied]
= $(Y_N - Y_0/F_N)$

where Y_0 is the crop yield (kg ha^{-1}) measured in a treatment with no N application.

RE_N = apparent recovery efficiency of applied N (kg N taken up kg^{-1} N applied) = $(U_N - U_0)/F_N$

where U_N is the plant N uptake (kg ha^{-1}) measured in aboveground biomass at physiological maturity in a plot that received N at the rate of F_N (kg ha^{-1}) and U_0 is the N uptake measured in aboveground biomass in a plot without the addition of N.

PE_N = physiological efficiency of applied N (kg product increase kg^{-1} fertilizer N taken up) = $(Y_N - Y_0)/(U_N - U_0)$

In field studies, these different agronomic indices of NUE are calculated either from differences in aboveground biomass and N uptake between fertilized plots and an unfertilized control ("difference method") or from ^{15}N-labeled fertilizers.

Source: Cassman, K. G., S. Peng, D. C. Olk, J. K. Ladha, W. Reichardt, A. Dobermann, and U. Singh. 1998. Opportunities for increased nitrogen use efficiency from improved resource management in irrigated rice systems. *Field Crops Research* 56:7–38.

List of Contributors

Bruno J. R. Alves
Embrapa Agrobiologia
Km 47 ant. Rod. Rio-Sao Paulo
Seropedica 23.851-970
Cx Postal 74.505
Rio de Janeiro, Brazil

John F. Angus
CSIRO
Plant Industry
GPO Box 1600
Canberra, ACT 2601, Australia

Milka S. Aulakh
Department of Soils
Punjab Agricultural University
Ludhiana 141004
Punjab, India

Vethaiya Balasubramanian
Training, Delivery and Impact
IRRI - Los Banos
DAPO Box 7777
Metro Manila, The Philippines

Mateete Bekunda
Faculty of Agriculture
Makerere University
P.O. Box 7062
Kampala, Uganda

J. Berntsen
Department of Crop Physiology
    and Soil Science
P.O. Box 50
DK-8830 Tjele, Denmark

Elizabeth Boyer
College of Environmental Science
    and Forestry
State University of New York
204 Marshall Hall
1 Forestry Drive
Syracuse, NY 13210, USA

Roland J. Buresh
Crop, Soil and Water Sciences
    Division
IRRI
DAPO Box 7777
Metro Manila, The Philippines

Zucong Cai
Institute of Soil Science
Chinese Academy of Sciences
Nanjing 210008, China

Kenneth G. Cassman
Department of Agronomy
 and Horticulture
University of Nebraska-Lincoln
P.O. Box 830915
Lincoln, NE 68583-0915, USA

Phillip Chalk
IAEA
Wagramerstrasse 5
A-1400 Vienna, Austria

Brian S. Dear
NSW Agriculture
Wagga Wagga, NSW, Australia

Achim Dobermann
Department of Agronomy
 and Horticulture
University of Nebraska-Lincoln
253 Keim Hall
P. O. Box 830915
Lincoln, NE 68583-0915, USA

Laurie Drinkwater
Department of Horticulture
Cornell University
Ithaca, NY 14853, USA

J. Eriksen
Department of Crop Physiology
 and Soil Science
P.O. Box 50
DK-8830 Tjele, Denmark

John R. Freney
CSIRO
62 Gellibrand St.
Campbell, ACT 2612, Australia

James N. Galloway
Department of Environmental
 Sciences
University of Virginia—Clark Hall
291 McCormack Road
P.O. Box 400123
Charlottesville, VA 22904-4123, USA

Kenneth E. Giller
Department of Plant Sciences
Building 537, Haarweg 333
Wageningen University
P.O. Box 430
6700 AK Wageningen, The
 Netherlands

Susan Greenwood Etienne
SCOPE
51, bd de Montmorency
75016 Paris, France

Keith Goulding
Agriculture and the Environment
 Division
Rothamsted Research
Harpenden
Hertfordshire AL5 2JQ, UK

Lawrence L. Hammond
Resource Development Division
International Fertilizer Development
 Center
P.O. Box 2040
Muscle Shoals, AL 35662, USA

H. Hauggaard-Nielsen
Risø National Laboratory
Roskilde, Denmark

John Havlin
North Carolina State University
2234 Williams Hall Campus
Box 7619
Raleigh, NC 27695-7619, USA

Patrick Heffer
International Fertilizer Industry
  Association
28, rue Marbeuf
75008 Paris, France

Julio Henao
International Fertilizer Development
  Center
P.O. Box 2040
Muscle Shoals, AL 35662, USA

Robert W. Howarth
Corson Hall, Cornell University
Ithaca, NY 14853, USA

Jian Liang Huang
College of Biological Sciences and
  Rice Research Institute
Hunan Agricultural University
Changsha, Hunan 410128, China

E. S. Jensen
Risø National Laboratory
Roskilde, Denmark

T. J. Krupnik
International Agricultural
  Development
University of California
One Shields Avenue
Davis, CA 95616-8515, USA

Jagdish. K. Ladha
IRRI
Los Banos, The Philippines

Pedro L. O. de A. Machado
Embrapa Solos
Rua Jardim Botanico 1024
22460-000 Rio de Janeiro, RJ, Brazil

Luc Maene
International Fertilizer Industry
  Association
28 rue Marbeuf
75008 Paris, France

Tariq Mahmood
Soil Biology & Plant Nutrition
  Division
Nuclear Institute for Agriculture
  & Biology
P.O. Box 128, Jhang Road
Faisalabad 38000, Pakistan

Sukhdev S. Malhi
Agriculture and Agri-Food Canada
Research Farm
P.O. Box 1240
Melfort, Saskatchewan, Canada S0E
  1A0

Jerry M. Melillo
The Ecosystems Center
Marine Biological Laboratory
7 MBL St.
Woods Hole, MA 02543, USA

R. Merckx
Laboratory for Soil and Water
 Management
Faculty of Agricultural and Applied
 Biological Sciences
K.U. Leuven, Kasteelpark Arenberg 20
3001 Heverlee, Belgium

Arvin R. Mosier
USDA/ARS
2150 Centre Avenue
Building D, Suite 100
Fort Collins, CO 80526, USA

Daniel Mugendi
Kenyatta University
Faculty of Environmental Studies
P.O. Box 43844
Nairobi, Kenya

Scott Murrell
Potash & Phosphate Institute
3579 Commonwealth Road
Woodbury, MN 55125, USA

Justice Nyamangara
Department of Soil Science &
 Agricultural Engineering
University of Zimbabwe
MP167, Mount Pleasant
Harare, Zimbabwe

Phibion Nyamudeza
Price Waterhouse Coopers
Building No. 4, Arundel Office Park
Norfolk Road, Mt. Pleasant
P. O. Box 453
Harare, Zimbabwe

Victor A. Ochwoch
Department of Soil Science
Makerere University
P.O. Box 7062
Kampala, Uganda

Oene Oenema
Wageningen University
6700 AK Wageningen, The
 Netherlands

Jorgen E. Olesen
Department of Crop Physiology
 and Soil Science
P.O. Box 50
DK-8830 Tjele, Denmark

M. J. Paine
International Agricultural
 Development
University of California
One Shields Avenue
Davis, CA 95616-8515, USA

Cheryl Palm
Senior Research Scientist
The Earth Institute at Columbia
 University
117 Monell Building
P.O. Box 1000
Palisades, NY 10964-8000, USA

John Pender
International Food Policy Research
 Institute
2033 K St. N.W.
Washington, D.C. 20006, USA

Shaobing Peng
Crop, Soil and Water Sciences
  Division
IRRI
DAPO Box 7777
Metro Manila, The Philippines

Mark B. Peoples
CSIRO
Plant Industry
GPO Box 1600
Canberra, ACT 2601, Australia

Véronique Plocq-Fichelet
SCOPE
51, bd de Montmorency
75016 Paris, France

Mark Rosegrant
International Food Policy Research
  Institute
2033 K. St., N.W.
Washington, D.C. 20006, USA

Thomas Rosswall
ICSU
51, bd de Montmorency
75016 Paris, France

Amit H. Roy
Resource Development Division
IFDC
P. O. Box 2040
Muscle Shoals, AL 35662, USA

Megan. H. Ryan
The University of Western Australia
Crawley, WA, Australia

N. Sanginga
Tropical Soil Biology and Fertility
  Institute of the International Centre
  for Tropical Agriculture
P.O. Box 30677
Nairobi, Kenya

Mary Scholes
Department of Animal, Plant and
  Environmental Sciences
University of the Witwatersrand
Johannesburg
Private Bag 3
Wits 2050, South Africa

Elsje L. Sisworo
Centre for Application of Isotopes
  and Radiation
Jalan Cinere Pasar Jumat Kotak
P.O. Box 7002 JKSKL
Jakarta 12070, Indonesia

Johan Six
Department of Agronomy and
  Range Science
University of California
One Shields Avenue
Davis, CA 95616-8515, USA

P. Sørensen
Department of Crop Physiology
  and Soil Science
P.O. Box 50
DK-8830 Tjele, Denmark

Henry Ssali
Kawanda Agricultural Research
  Institute
P.O. Box 7065
Kampala, Uganda

John W. B. Stewart
SCOPE Editor-in-Chief
118 Epron Road
Salt Spring Island, B.C., Canada V8K
1C7

J. Keith Syers
Naresuan University
Phitsanulok 65000, Thailand

A. D. Swan
CSIRO
Plant Industry
GPO Box 1600
Canberra, ACT 2601, Australia

A. G. Thomsen
Department of Crop Physiology
and Soil Science
P.O. Box 50
DK-8830 Tjele, Denmark

J. K. Thomsen
Department of Crop Physiology
and Soil Science
P.O. Box 50
DK-8830 Tjele, Denmark

Oswald Van Cleemput
Ghent University
Faculty of Agriculture and Applied
Biological Sciences
Coupure 653
B-9000 Ghent, Belgium

Chris Van Kessel
Department of Agronomy and Range
Sciences
University of California-Davis
One Shield Avenue
Davis, CA 95616-8515, USA

Bernard Vanlauwe
TSBF-CIAT
P.O. Box 30677
Nairobi, Kenya

J. Virgona
Charles Sturt University
Wagga Wagga, NSW, Australia

Guanghuo Wang
College of Environmental and
Natural Resources Science
Zhejiang University
Hangzhou, Zhejiang 310029, China

Stanley Wood
Environment and Production
Technology Division
IFPRI
Washington, D.C. 20006, USA

Kazuyuki Yagi
Greenhouse Gas Team
National Institute for Agro-
Environmental Sciences
3-1-3 Kannondai
Tsukuba, Ibaraki 305-8604, Japan

Jianchang Yang
Agronomy Department
Agricultural College
Yangzhou University
Yangzhou, Jiangsu 225009, China

Xuhua Zhong
Rice Research Institute
Guangdong Academy of Agricultural
  Science
Guangzhou, Guangdong 510640,
  China

Yingbin Zhou
College of Biological Sciences and
  Rice Research Institute
Hunan Agricultural University
Changsha, Hunan 410128, China

SCOPE Series List

SCOPE 1–59 are now out of print. Selected titles from this series can be downloaded free of charge from the SCOPE Web site (http://www.icsu-scope.org).

SCOPE 1: *Global Environment Monitoring*, 1971, 68 pp
SCOPE 2: *Man-made Lakes as Modified Ecosystems*, 1972, 76 pp
SCOPE 3: *Global Environmental Monitoring Systems (GEMS): Action Plan for Phase I*, 1973, 132 pp
SCOPE 4: *Environmental Sciences in Developing Countries*, 1974, 72 pp
SCOPE 5: *Environmental Impact Assessment: Principles and Procedures*, Second Edition, 1979, 208 pp
SCOPE 6: *Environmental Pollutants: Selected Analytical Methods*, 1975, 277 pp
SCOPE 7: *Nitrogen, Phosphorus and Sulphur: Global Cycles*, 1975, 129 pp
SCOPE 8: *Risk Assessment of Environmental Hazard*, 1978, 132 pp
SCOPE 9: *Simulation Modelling of Environmental Problems*, 1978, 128 pp
SCOPE 10: *Environmental Issues*, 1977, 242 pp
SCOPE 11: *Shelter Provision in Developing Countries*, 1978, 112 pp
SCOPE 12: *Principles of Ecotoxicology*, 1978, 372 pp
SCOPE 13: *The Global Carbon Cycle*, 1979, 491 pp
SCOPE 14: *Saharan Dust: Mobilization, Transport, Deposition*, 1979, 320 pp
SCOPE 15: *Environmental Risk Assessment*, 1980, 176 pp
SCOPE 16: *Carbon Cycle Modelling*, 1981, 404 pp
SCOPE 17: *Some Perspectives of the Major Biogeochemical Cycles*, 1981, 175 pp
SCOPE 18: *The Role of Fire in Northern Circumpolar Ecosystems*, 1983, 344 pp
SCOPE 19: *The Global Biogeochemical Sulphur Cycle*, 1983, 495 pp
SCOPE 20: *Methods for Assessing the Effects of Chemicals on Reproductive Functions*, SGOMSEC 1, 1983, 568 pp
SCOPE 21: *The Major Biogeochemical Cycles and their Interactions*, 1983, 554 pp
SCOPE 22: *Effects of Pollutants at the Ecosystem Level*, 1984, 460 pp

SCOPE 23: *The Role of Terrestrial Vegetation in the Global Carbon Cycle: Measurement by Remote Sensing,* 1984, 272 pp

SCOPE 24: *Noise Pollution,* 1986, 466 pp

SCOPE 25: *Appraisal of Tests to Predict the Environmental Behaviour of Chemicals,* 1985, 400 pp

SCOPE 26: *Methods for Estimating Risks of Chemical Injury: Human and Non-Human Biota and Ecosystems, SGOMSEC 2,* 1985, 712 pp

SCOPE 27: *Climate Impact Assessment: Studies of the Interaction of Climate and Society,* 1985, 650 pp

SCOPE 28: *Environmental Consequences of Nuclear War*
Volume I: Physical and Atmospheric Effects, 1986, 400 pp
Volume II: Ecological and Agricultural Effects, 1985, 563 pp

SCOPE 29: *The Greenhouse Effect, Climatic Change and Ecosystems,* 1986, 574 pp

SCOPE 30: *Methods for Assessing the Effects of Mixtures of Chemicals, SGOM-SEC 3,* 1987, 928 pp

SCOPE 31: *Lead, Mercury, Cadmium and Arsenic in the Environment,* 1987, 384 pp

SCOPE 32: *Land Transformation in Agriculture,* 1987, 552 pp

SCOPE 33: *Nitrogen Cycling in Coastal Marine Environments,* 1988, 478 pp

SCOPE 34: *Practitioner's Handbook on the Modelling of Dynamic Change in Ecosystems,* 1988, 196 pp

SCOPE 35: *Scales and Global Change: Spatial and Temporal Variability in Biospheric and Geospheric Processes,* 1988, 376 pp

SCOPE 36: *Acidification in Tropical Countries,* 1988, 424 pp

SCOPE 37: *Biological Invasions: a Global Perspective,* 1989, 528 pp

SCOPE 38: *Ecotoxicology and Climate with Special References to Hot and Cold Climates,* 1989, 432 pp

SCOPE 39: *Evolution of the Global Biogeochemical Sulphur Cycle,* 1989, 224 pp

SCOPE 40: *Methods for Assessing and Reducing Injury from Chemical Accidents, SGOMSEC 6,* 1989, 320 pp

SCOPE 41: *Short-Term Toxicity Tests for Non-genotoxic Effects, SGOMSEC 4,* 1990, 353 pp

SCOPE 42: *Biogeochemistry of Major World Rivers,* 1991, 356 pp

SCOPE 43: *Stable Isotopes: Natural and Anthropogenic Sulphur in the Environment,* 1991, 472 pp

SCOPE 44: *Introduction of Genetically Modified Organisms into the Environment,* 1990, 224 pp

SCOPE 45: *Ecosystem Experiments,* 1991, 296 pp

SCOPE 46: *Methods for Assessing Exposure of Human and Non-human Biota SGOMSEC 5,* 1991, 448 pp

SCOPE 47: *Long-Term Ecological Research. An International Perspective*, 1991, 312 pp

SCOPE 48: *Sulphur Cycling on the Continents: Wetlands, Terrestrial Ecosystems and Associated Water Bodies*, 1992, 345 pp

SCOPE 49: *Methods to Assess Adverse Effects of Pesticides on Non-target Organisms, SGOMSEC 7*, 1992, 264 pp

SCOPE 50: *Radioecology after Chernobyl*, 1993, 367 pp

SCOPE 51: *Biogeochemistry of Small Catchments: a Tool for Environmental Research*, 1993, 432 pp

SCOPE 52: *Methods to Assess DNA Damage and Repair: Interspecies Comparisons, SGOMSEC 8*, 1994, 257 pp

SCOPE 53: *Methods to Assess the Effects of Chemicals on Ecosystems, SGOMSEC 10*, 1995, 440 pp

SCOPE 54: *Phosphorus in the Global Environment: Transfers, Cycles and Management*, 1995, 480 pp

SCOPE 55: *Functional Roles of Biodiversity: a Global Perspective*, 1996, 496 pp

SCOPE 56: *Global Change, Effects on Coniferous Forests and Grasslands*, 1996, 480 pp

SCOPE 57: *Particle Flux in the Ocean*, 1996, 396 pp

SCOPE 58: *Sustainability Indicators: a Report on the Project on Indicators of Sustainable Development*, 1997, 440 pp

SCOPE 59: *Nuclear Test Explosions: Environmental and Human Impacts*, 1999, 304 pp

SCOPE 60: *Resilience and the Behavior of Large-Scale Systems, 2002, 287 pp*

SCOPE 61: *Interactions of the Major Biogeochemical Cycles: Global Change and Human Impacts*, 2003, 384 pp

SCOPE 62: *The Global Carbon Cycle: Integrating Humans, Climate, and the Natural World*, 2004, 526 pp

SCOPE 63: *Alien Invasive Species: A New Synthesis*, 2004, 352 pp.

SCOPE 64: *Sustaining Biodiversity and Ecosystem Services in Soils and Sediments*, 2003, 308 pp

SCOPE Executive Committee 2001–2004

President:
Dr. Jerry M. Melillo (USA)

1st Vice-President
Prof. Rusong Wang (China-CAST)

2nd Vice-President
Prof. Bernard Goldstein (USA)

Treasurer
Prof. Ian Douglas (UK)

Secretary-General
Prof. Osvaldo Sala (Argentina-IGBP)

Members
Prof. Himansu Baijnath (South Africa-IUBS)
Prof. Manuwadi Hungspreugs (Thailand)
Prof. Venugopalan Ittekkot (Germany)
Prof. Holm Tiessen (Canada)
Prof. Reynaldo Victoria (Brazil)

Index

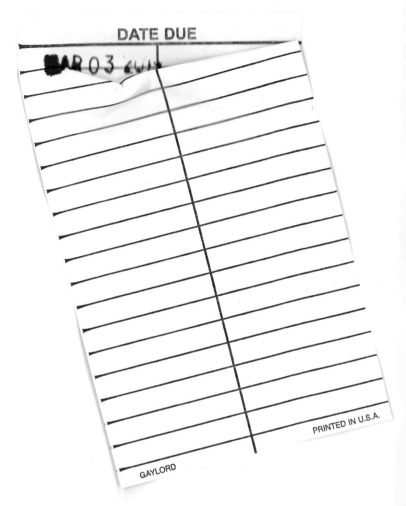

DATE DUE

MAR 03 20..

PRINTED IN U.S.A.

GAYLORD